The Unnatural Nature of Science

THE UNNATURAL NATURE
OF SCIENCE

LEWIS WOLPERT

HARVARD UNIVERSITY PRESS
CAMBRIDGE, MASSACHUSETTS

Lewis Wolpert
Copyright © 1992
All rights reserved
Printed in the United States of America
Second printing, 1994

Published by agreement with
Faber and Faber Limited, London

First Harvard University Press paperback edition, 1994

Library of Congress Cataloging-in-Publication Data
Wolpert, L. (Lewis)
    The unnatural nature of science / Lewis Wolpert.
      p.   cm.
    Includes bibliographical references and index.
    ISBN 0-674-92980-2 (cloth)
    ISBN 0-674-92981-0 (pbk.)
    1. Science—Philosophy.   2. Science—Social aspects.   I. Title.
Q175.W737   1993   92-40510
501—dc20   CIP

Contents

Preface

This book has its origin in dissatisfaction and a puzzle. The dissatisfaction is with the public image of science and with much of the writing about science in the media as well as that by academics including philosophers and sociologists. The puzzle is why the nature of science should be so misunderstood and why non-scientists have so much difficulty understanding scientific ideas. This lack of understanding seemed to be linked to a certain fear of and even hostility to science itself.

So I have tried to present science in a new light, which I hope will help to resolve some of these problems. By dealing with so broad a topic as the nature of science, I have inevitably touched on areas in which I have no formal training such as philosophy, psychology and history. I am by profession a research biologist in the field of embryology, and my approach can best be characterized as that used in natural history. I have therefore sought much advice, some of which is acknowledged below, and I am very grateful to everyone who has helped me. I am also indebted to Warwick University for inviting me to give the 1990 Radcliffe Lectures on 'Science: An Unnatural History', which laid the foundations for what is presented here.

I thank Percy Cohen, Stephen Cang, Patricia Farrar, Christopher Gardner, Jonathan Glover, Mary and Jack Herberg, Judy Hicklin, Frank James, Jonckheere, Roger Jowell, Michael Kidron, Roland Littlewood, Lauro Martines, Arthur Miller, Timothy McDermott and Mary Tuck.

Maureen Maloney needs special thanks for her patience in preparing the manuscript.

I am also specially indebted to my editors, Bob Davenport and, foremost, Susanne McDadd.

Introduction

Knowledge has killed the sun, making it a ball of gas
with spots . . . The world of reason and science . . .
this is the dry and sterile world the abstracted mind
inhabits.   *D. H. Lawrence*

Modern science . . . abolishes as mere fiction the
innermost foundations of our natural world: it kills
God and takes his place on the vacant throne so
henceforth it would be science that would hold the
order of being in its hand as its sole legitimate
guardian and so be the legitimate arbiter of all relevant
truth . . . People thought they could explain and
conquer nature – yet the outcome is that they
destroyed it and disinherited themselves from it.
                                            Václav Havel

A public that does not understand how science works
can, all too easily, fall prey to those ignoramuses . . .
who make fun of what they do not understand, or
to the sloganeers who proclaim scientists to be the
mercenary warriors of today, and the tools of the
military. The difference . . . between . . .
understanding and not understanding . . . is also the
difference between respect and admiration on the one
side, and hate and fear on the other.   *Isaac Asimov*

Science is arguably the defining feature of our age; it characterizes Western civilization. Science has never been more
successful nor its impact on our lives greater, yet the ideas
of science are alien to most people's thoughts. It is striking
that about half the population of the United States does not
believe in evolution by natural selection and that a significant
proportion of British citizens does not think the earth goes
round the sun. And I doubt that of those who do believe the
earth moves round the sun, even one person in 100,000 could

give sound reasons for their conviction (the evidence and the arguments for such a belief are in fact quite complex). Indeed, many people accept the ideas of science because they have been told that these ideas are true rather than because they understand them. No wonder the nature of science is so poorly understood. Instead it is viewed with a mixture of admiration and fear, hope and despair, seen both as the source of many of the ills of modern industrial society and as the source from which cures for these ills will come.

Some of the anti-science attitudes are not new: Mary Shelley's Dr Frankenstein, H. G. Wells's Dr Moreau and Aldous Huxley's *Brave New World*, for example, are evidence of a powerfully emotive anti-science movement. Science is dangerous, so the message goes – it dehumanizes; it takes away free will; it is materialistic and arrogant. It removes magic from the world and makes it prosaic. But note where these ideas come from – not from the evidence of history, but from creative artists who have moulded science by their own imagination. It was Mary Shelley who created Frankenstein's monster, not science, but its image is so powerful that it has fuelled fears about genetic engineering that are very hard to remove.

Current attitudes to science indicate both ambivalence and polarization. Surveys confirm that there is much interest in, and admiration for, science, coupled with an unrealistic belief that it can cure all problems; but there is also, for some, a deep-seated fear and hostility, with several lines of criticism. Science is perceived as materialist and as destructive of any sense of spiritual purpose or awareness; it is held responsible for the threat of nuclear warfare and for the general disenchantment with a modern industrial society that pollutes and dehumanizes. The practitioners of science are seen as cold, anonymous and uncaring technicians. The fear of genetic engineering and the manipulation of embryos looms large, and the image of Dr Frankenstein is increasingly embellished. The image of scientists themselves remains as stereotyped and inaccurate as ever: when not crazy, they appear bedecked in a white coat, wearing spectacles, and wielding a test-tube. The media usually present scientists as totally anonymous and character-free and give little insight into the way in which they work. Scientists are still widely perceived as being like Mr Gradgrind in Charles Dickens's *Hard Times*, interested only in facts and yet more facts, the collection of which is the hallmark of the scientific enterprise, and the overwhelming burden of which seems to drive them into increasingly obscure specializations. Almost as misleading is the idea that there is a 'scientific method' that provides a formula which, if faithfully followed, will lead to discovery. Any idea of creativity in science – which is rare – is linked, romantically and falsely, with that of artistic creativity.

Thirty years ago, C. P. Snow suggested that there were two separate cultures: one relating to science and the other to the arts and humanities. He was criticized for his use of the term 'culture'. Some people even argue that science is not part of culture at all: following Nietzsche's claim that science, with its reductionism and materialism, has deprived man of his special status, it seems to some that only an idea of culture that actually excludes science can restore man's dignity. Whatever the definition of culture, however, Snow was right in emphasizing that the 'culture' of science was different. What he did not do was to give any insight into why this should be.

Some of the hostility to science may be explained by the American literary critic Lionel Trilling's comment on the difficulty non-scientists have in understanding science: 'This exclusion of most of us from the mode of thought which is habitually said to be the characteristic achievement of the modern age is bound to be experienced as a wound to our intellectual self-esteem.'

The central theme presented in this book is that many of the misunderstandings about the nature of science might be corrected once it is realized just how 'unnatural' science is. I will argue that science involves a special mode of thought and is unnatural for two main reasons, which are developed in Chapter 1. Firstly, the world just is not constructed on a common-sensical basis. This means that 'natural' thinking – ordinary, day-to-day common sense – will never give an understanding about the nature of science. Scientific ideas are, with rare exceptions, counter-intuitive: they cannot be acquired by simple inspection of phenomena and are often outside everyday experience. Secondly, doing science requires a conscious awareness of the pitfalls of 'natural' thinking. For common sense is prone to error when applied to problems requiring rigorous and quantitative thinking; lay theories are highly unreliable.

In establishing the unnatural nature of science, it is essential to distinguish between science and technology, particularly since the two are so often confused. The evidence for the distinction, discussed in Chapter 2, comes largely from history. Technology is very much older than science, and most of its achievements – from primitive agriculture to the building of great churches and the invention of the steam engine – have in no way been dependent on science. Even the mode of thought in technology is very different from that of science.

Once the distinction between science and technology is recognized then the origins of science in Greece take on a special significance, which is the subject of Chapter 3. The peculiar nature of science is responsible for science having arisen only once. Even though most, if not all, of Aristotle's science was wrong – he can be thought of as the scientist of common sense – he established the basis of a system for explaining the world based on postulates and logical deduction. This was brilliantly exploited by Euclid and Archimedes. By contrast the Chinese, often thought of as scientists, were expert engineers but made negligible contributions to science. Their philosophies were essentially mystical, and it may have been rationality and a concept of laws governing nature that allowed science to develop in the West.

Since science is unique, it is to be expected that scientific creativity has its own special characteristics quite different from those of the arts, as we shall see in Chapter 4. Scientific genius is often characterized by a 'psychic courage' which requires scientists to include in their ideas assumptions for which they have very little evidence. Scientific creativity is, of course, not understood, and one should be sceptical both of the suggestion that it involves merely a sort of problem-solving that can be done by computers and of the theory that it is heavily dependent on chance, characterized under the rubric of serendipity.

Because any scientific discovery can be made only once, scientific research generates intense competition, even though in the long term most scientists are anonymous, or their names are recorded only in a historical context. But the essential social nature of science, discussed in Chapter 5, engenders cooperation too. New ideas have to be accepted by consensus of the scientific community – and because there is often a reluctance to surrender current views, scientists may be unwise to abandon their ideas at the first indication they have been falsified. Scientists also judge theories on their explanatory value, simplicity and fruitfulness.

It might be thought that either philosophers or sociologists would have been able to illuminate the nature of science and why it has been so successful. Alas, not only have they failed to do so but some have instead provided what they regard as good reasons for doubting whether science really does provide an understanding of the way in which the world works, as we shall see in Chapter 6. Fortunately for science, these philosophical claims have no relevance to science and can be ignored. There are numerous 'styles' for doing science: the only constant is the need to measure one's ideas against the real world.

But it must be admitted that it is not always easy to explain the confidence with which one can distinguish science from non-science. One approach, discussed in Chapter 7, is to recognize that some areas are premature or too primitive for scientific investigation. Just as in the seventeenth and eighteenth centuries the great debate about the nature of the development of the embryo – whether all organs were preformed or actually were made during development – could not be resolved until other advances in biology had been made, so the claims made for the scientific nature of psychoanalysis may be premature given the current state of knowledge about the brain, particularly since the mechanisms that psychoanalysis proposes are little different from the phenomena they attempt to explain, as was also the case in early embryology. Claims for paranormal phenomena are easily dealt with because the evidence is so poor, but a special problem is raised by religion: while religious belief is incompatible with science, many scientists are deeply religious. An explanation of this paradox is the difference between natural and unnatural thinking.

There remains the major problem that scientific knowledge is perceived as being dangerous. Was it not responsible for nuclear warfare and the current unease about genetic engineering? Using the history of the atomic bomb and of eugenics as examples, Chapter 8 discusses the social obligations of science and argues that many of the so-called new ethical problems are merely reflections of a failure to understand the nature of science.

While science provides our best hope for solving many major problems such as environmental pollution and genetic diseases, it does have its limits, and these and the need for a more accurate public perception of science's nature and processes are discussed in Chapter 9.

Science can be quite uncomfortable to live with – at least for some people. It offers no hope for an afterlife, it tolerates no magic and it doesn't tell us how to live. But there is no good reason to believe, with D. H. Lawrence, that scientific understanding creates a 'dry and sterile world' by apparently removing all mystery. To quote Einstein, 'the greatest mystery of all is the (partial) intelligibility of the world.' And science itself can be very beautiful.

I
Unnatural Thoughts

P 1-24

I.

It is often held that science and common sense are closely
linked. Thomas Henry Huxley, Darwin's brilliant colleague,
spoke of science as being nothing more than trained common
sense. 'Science is rooted in the whole apparatus of common-
sense thought' was the optimistic claim of the philosopher
and mathematician Alfred North Whitehead. However
reasonable they may sound, such views are, alas, quite misleading. In fact, both the ideas that science generates and the
way in which science is carried out are entirely counter-
intuitive and against common sense – by which I mean that
scientific ideas cannot be acquired by simple inspection of
phenomena and that they are very often outside everyday
experience. Science does not fit with our natural expectations.

Common sense is not a simple thing: it reflects an enormous amount of information that one has gained about the
world and provides a large number of practical rules – many
of them quite logical – for dealing with day-to-day life. It is
so much a part of everyday life that one seldom thinks about
it. It will be considered shortly.

An immediate problem in comparing common sense with
science is, of course, defining what is meant by 'science'.
Providing a rigorous definition is far from easy, and the best
way to advance at this stage is by example.

Physics is probably a good way of showing what is meant
by science: it tries to provide an explanation of nature – the
world we live in – at the most fundamental level. It aims to
find explanations for an enormous variety of phenomena –
the movement of all objects; the nature of light and sound,
heat and electricity; the fundamental constitution of matter
– in terms of as few principles as possible. Rigorous theories

are constructed which explain observed phenomena, and these theories must be capable of being tested by both confirmation and attempts to falsify them. It is also an absolute requirement that theories must be capable of modification, or even abandonment, when evidence demands it. In this process, all the phenomena must be capable of observation by independent observers, for scientific knowledge is public knowledge.

Science always relates to the outside world, and its success depends on how well its theories correspond with reality. Criteria for a good theory – in addition to explaining observations and predicting new ones – include relative simplicity and elegance, and as scientists themselves repeatedly point out, a good theory should raise interesting new questions.

For Einstein, the object of all science was 'to coordinate our experiments and bring them into a logical system'. In this endeavour, mathematics plays a fundamental role for expressing scientific ideas in quantitative terms: for the nineteenth-century physicist Lord Kelvin, one could only really claim to know something if one could measure what one was speaking about and express it in numbers. While his was an extreme view, and can certainly be shown to be wrong, the attempt to express ideas with mathematical rigour underlies much of scientific endeavour. Newton's laws of motion provide a wonderful triumph of this approach: with a few basic laws of motion together with mathematics it is possible to explain an enormous range of movements – from those of the planets to those of billiard and tennis balls.

The physics of motion provides one of the clearest examples of the counter-intuitive and unexpected nature of science. Most people not trained in physics have some sort of vague ideas about motion and use these to predict how an object will move. For example, when students are presented with problems requiring them to predict where an object – a bomb, say – will land if dropped from an aircraft, they often get the answer wrong. The correct answer – that the bomb will hit that point on the ground more or less directly below the point at which the aircraft has arrived at the moment of impact – is often rejected. The underlying confusion partly comes from not recognizing that the bomb continues to move forward when released and this is not affected by its downwards fall. This point is made even more dramatically by another example. Imagine being in the centre of a very large flat field. If one bullet is dropped from your hand and another is fired horizontally from a gun at exactly the same time, which will hit the ground first? They will, in fact, hit the ground at the same time, because the bullet's rate of fall is quite independent of its horizontal motion. That the bullet which is fired is travelling horizontally has no effect on how fast it falls under the action of gravity.

Another surprising feature of motion is that the most natural state for an object is movement at constant speed – not, as most of us think, being stationary. A body in motion will continue to move forever unless there is a force that stops it. This was a revolutionary idea first proposed by Galileo in the early seventeenth century and was quite different from Aristotle's more common-sense view, from the fourth century BC, that the motion of an object required the continuous action of a force. Galileo's argument is as follows. Imagine a perfectly flat plane and a perfectly round ball. If the plane is slightly inclined the ball will roll down it and go on and on and on. But a ball going up a slope with a slight incline will have its velocity retarded. From this it follows that motion along a horizontal plane is perpetual, 'for if the velocity be uniform it cannot be diminished or slackened, much less destroyed.' So, on a flat slope, with no resistance, an initial impetus will keep the ball moving forever, even though there is no force. Thus the natural state of a physical object is motion along a straight line at constant speed, and this has come to be known as Newton's first law of motion. That a real ball will in fact stop is due to the opposing force provided by friction between a real ball and a real plane. The enormous conceptual change that the thinking of Galileo required shows that science is not just about accounting for the 'unfamiliar' in terms of the familiar. Quite the contrary: *science often explains the familiar in terms of the unfamiliar.*

Aristotle's idea of motion – that it requires the constant application of a force – is familiar to us in a way that Galileo's and Newton's never can be. So it is not surprising that, when asked to indicate the forces on a ball thrown up, many students imagine an upward force to be present after the ball leaves the hand, whereas the truth is that at all stages after the ball leaves the hand it experiences only a downward force due to gravity. This is no simple problem and even Galileo got it wrong, though he did recognize that there was a problem. Newton's second law provides the explanation. Forces acting on a body cause it to accelerate, so forces can either increase or decrease its speed. When a ball is thrown up, it would continue upwards forever if there were no forces like friction or gravity to slow it down. The force of gravity acts to accelerate the ball *towards* the earth – which is equivalent to a retardation in the ball's movement *away from* the earth – so the ball is slowed down and eventually reverses its upwards motion.

The naïve views held by the students are very similar to the 'impetus' theory put forward by Philoponus in the sixth century and by John Buridan in the fourteenth century. This theory assumes that the act of setting an object in motion impresses on that object a force or impetus that keeps it in motion. Persistence of thinking in terms of impetus over the three hundred years since Newton shows how difficult it is to assimilate a counter-intuitive scientific idea.

The nature of white light is another counter-intuitive example from physics which was also discovered by Newton. Newton showed that ordinary white light is a mixture of different kinds of light, each of which we see as coloured. When all the colours of the rainbow are combined, the result is white.

Yet another example is provided by the phlogiston theory in the eighteenth century, which addressed the problem of what happens when an object burns. In Aristotelian terms, and common sense, when anything burns, something clearly leaves the burning object. This something was thought to be phlogiston. Again common sense is misleading, for an essential feature of burning is that oxygen is taken up rather than something being released.

Even something as simple as the mechanism involved in the spread of a dye in water does not accord with common sense. Consider placing a drop of ink, or a dye, at one end of a trough of water. In time, the dye will spread across all of the water. Why does it spread? It might seem that there is something about the high concentration at one end 'driving' the dye away. In fact, on the contrary, the spread is all due to the random motion of the dye molecules; if one could follow the movement of any single molecule, one would not be able to determine the direction in which the dye spreads. Again, is it intuitive that temperature, hot and cold, reflects a similar underlying property related to the vibration of molecules?

Science also deals with enormous differences in scale and time compared with everyday experience. Molecules, for example, are so small that it is not easy to imagine them. If one took a glass of water, each of whose molecules were tagged in some way, went down to the sea, completely emptied the glass, allowed the water to disperse through all the oceans, and then filled the glass from the sea, then almost certainly some of the original water molecules would be found in the glass. What this means is that there are many more molecules in a glass of water than there are glasses of water in the sea. There are also, to give another example, more cells in one finger than there are people in the world. Again, geological time is so vast – millions and millions of years – that it was one of the triumphs of nineteenth-century geology to recognize that the great mountain ranges, deep ravines and valleys could be accounted for by the operation of forces no different from those operating at present but operating over enormous periods of time. It was not necessary to postulate catastrophes.

A further example of where intuition usually fails, probably because of the scale, is provided by imagining a smooth globe as big as the earth, round whose equator – 25,000 miles long – is a string that just fits. If the length of the string is increased by 36 inches, how far from the surface of the globe will the string stand out? The answer is about 6 inches, and is independent of whether the globe's equator is 25,000 or 25 million miles long.

There are rare exceptions to the rule that all scientific ideas are contrary to common sense. Ohm's law is the best example: the greater the resistance of an electric circuit, the greater is the voltage required to drive a current through the circuit. This does accord with everyday expectation. Generally, however, the way in which nature has been put together and the laws that govern its behaviour bear no apparent relation to everyday life. The laws of nature just cannot be inferred from normal day-to-day experience. Even that the earth goes round the sun is accepted more by authority than by genuine understanding – to provide the evidence is no trivial matter. As Bertrand Russell pointed out, we all start from a 'naïve realism', believing that things are what they seem. Thus we think that grass is green, that stones are hard and that snow is cold. But physics teaches us that the greenness of grass, the hardness of stones and the coldness of snow are not the greenness, hardness and coldness that we know in our own experience, but something very different. The same may even be true of economics – the Nobel laureate James Meade would like his tombstone to bear this epitaph: 'He tried to understand economics all his life, but common sense kept getting in the way.'

That science is an unnatural mode of thought, even disconcerting, was clearly understood by Aristotle:

In some ways, the effect of achieving understanding is to reverse completely our initial attitude of mind. For everyone starts (as we have said) by being perplexed by some fact or other: for instance . . . the fact that the diagonal of a square is incommensurable with the side. Anyone who has not yet seen why the side and the diagonal have no common unit regards this as quite extraordinary. But one ends up in the opposite frame of mind . . . for nothing would so much flabbergast a mathematician as if the diagonal and side of a square were to become commensurable.

Aristotle was referring to the fact that according to Pythagoras's theorem the diagonal of a square is a multiple of the square root of 2, and this is not a whole number but has as many figures after 1.4142 . . . as you would wish to calculate. But, in a way, to speak of the unnatural nature of scientific ideas is almost a circular statement: if scientific ideas were natural, they would not have required the difficult and protracted techniques of science for their discovery. All the examples so far refer to relatively simple scientific principles.

When one enters the world of cosmology, with its black holes and with the suggestion – or rather conviction – that the universe had its origin in a 'big bang' – that the universe was created in a few minutes in the distant past – then the science is beyond being counter-intuitive and becomes incomprehensible, or even magical, for those not trained in physics. Again, the world of the subatomic particles is full of ideas that have no correspondence to everyday life. This is a world where every electron – the smallest charged particle which is a constituent of all matter – is identical and where Heisenberg's uncertainty principle operates: there is no way of devising a method for pinpointing the exact position of the particles which make up atoms without sacrificing some precision about how fast they are moving. In this subatomic world, the rules for the behaviour of the subatomic particles are governed by quantum mechanics, in which the sort of causality we are familiar with no longer applies and the unpredictability of some events, such as radioactive decay, is a feature of the theory. Even Einstein could not accept this apparent lack of causality and the role of chance – hence his famous aphorism 'God does not play dice.'

It is one of the most unnatural features of science that the abstract language of mathematics should provide such a powerful tool for describing the behaviour of systems both inanimate, as in physics, and living, as in biology. Why the world should conform to mathematical descriptions is a deep question. Whatever the answer, it is astonishing.

Because so much of science is based on mathematics, it is not easy to explain scientific ideas in ordinary language. Moreover, understanding science is a hierarchical process: it is extremely difficult to understand the more advanced concepts until the basic concepts have been mastered. It is often even difficult to put the concepts into everyday language, particularly in physics, where mathematics plays a crucial role: there need not necessarily be a simple translation from mathematical formulations into concepts that make sense in terms of observable objects. It is this that makes quantum mechanics, black holes and much of physics inaccessible to most people. The same is also true of, for example, chemistry. Most chemical formulae which show the structural relations between the atoms, do not easily translate into common language. The formula for cholesterol, for example, conveys little to the non-chemist.

The basic concepts of molecular biology are no more intuitive than those of physics, and, since I will on several occasions use these concepts to illustrate ideas about science, it is necessary to describe some of them in a little detail.

That DNA is the genetic material – the physical basis of heredity – is quite well known. Involving no mathematics, its role is one of the easiest of the basic ideas of science to explain, yet it is really quite complex, and is built on a technical background. Even to recognize that there was something which might be identifiable as the genetic material required the work of a large number of scientists. People had long been aware that children resemble their parents, in both the human and animal world, but the nature of the mechanism which brought this about was not really understood until this century. Theories to explain this, from Aristotle onwards, included the idea of the transfer of some insubstantial 'pneuma' as the agent of inheritance, the idea that the father's contribution is the only significant factor, and the idea that the environment of the parents was a major determinant of the physical character of the offspring. Only towards the end of the last century did it become clear that chromosomes – string-like structures within the bag-like nucleus of the cell – could be the physical basis of heredity. It was only in the 1870s that the spermatozoa, which had first been seen under the microscope 200 years earlier, were at last recognized as being not parasites, as had been thought, but the means of providing the egg with the male genetic material. These discoveries required painstaking observation, ingenious experiments and technology such as microscopes. There was nothing in these basic discoveries that could have been expected from any normal experience of the world.

The identifying of DNA as the genetic material and of its role in controlling the behaviour of the cell required a further set of discoveries that were not based simply on biological experiments but also required quite complex physics and chemistry. Chemists had some time ago worked out the chemical composition of DNA – that it was made up essen-.

tially of four different smaller substances, or bases. But it was the discovery in 1953 of the structure of DNA – how the bases are arranged – which provided quite new insights. The structural analysis was based on X-ray diffraction, which is a technique used by physicists and chemists to obtain information about the three-dimensional arrangement of the atoms that are present in a molecule. An X-ray beam is shone through a crystal of the material and the rays that emerge give a complex, but characteristic, pattern of spots on a photographic plate. The spots reflect the way in which the X-rays were deflected as they passed through the crystal, and with skill and mathematical techniques it is possible, from these deflections, to work out the arrangement of the atoms.

James Watson and Francis Crick worked out the structure of DNA from both its chemical properties and X-ray diffraction. They required an enormous amount of background knowledge, and they worked very hard to get the answer. The result was a beautiful surprise, because it at once made clear one of the key features of life – replication. DNA is a very long string-like molecule made up of two strands twisted round each other in the form of a double helix. Each strand is made up of the four different bases, whose arrangement has two important features. Firstly, the bases are arranged in a very strict order unique to each individual along both strands, and the fundamental properties of DNA are determined by the particular sequence of bases. Secondly, the bases in one strand have a unique and complementary relationship to the bases in the other strand. Each base can only match with its complementary base. So, once the sequence of bases in one strand is specified, so too is the sequence of bases in the other strand. This provides the fundamental mechanism for replication: the two strands are unwound and then separated; then the cell synthesizes a complementary strand on each, by linking free bases (which are always present) to their complementary partner and then joining them up. This unexpected mechanism for the replication of DNA provides the essential basis for the replication of life itself, but this very simple description does no justice at all to the complex chemical events that are actually involved in the process.

The strict sequence of bases not only permits accurate replication but also provides the mechanism whereby DNA, as the genetic material, controls the behaviour of the cell. Cell behaviour is largely determined by a class of molecules called proteins. There are thousands of different proteins in the cell, and they are essential for all the key chemical reactions in the cell as well as providing the building blocks for the cell structures such as the filaments that generate muscle contractions. The character of a cell is entirely determined by which proteins it contains, and so the presence or absence of specific proteins controls cell behaviour.

DNA contains the code for all the proteins in the cell – a code in the sense that the sequence of the four bases in the DNA can be translated into a sequence of the twenty amino acids from which proteins are made. The properties of a protein are determined by the sequence of the string of amino acids from which it is made up. Thus the DNA of the cell is like a book which contains the recipe for every protein; the life of the cell, its character, is determined by which recipes are 'read', which proteins are made.

This elegant and universal mechanism is the basis for life, and on it rests all of biological science, genetics, cell biology, development and evolution. There was nothing in the day-to-day world to anticipate the ideas of modern cell and molecular biology. And there are two further implications which go quite against common sense: the failure of acquired characters to be inherited and the complete dependence of all evolutionary change on changes in the DNA. These two are, of course, intimately linked.

With very rare exceptions, all the characters that are passed from parents to offspring by ova and sperm are carried by the DNA. Any change in a character in an organism that can be inherited must involve a change in the DNA. This can result from different combinations of DNA being provided by the parents, and the sequence of the bases themselves may change due to mutations which are caused by errors at the time of replication or due to damage by environmental factors – by any process that changes the nature of the DNA so that the pattern of protein synthesis in some cells will be altered. Evolution is thus the continual change in the DNA of the cells from generation to generation. Between the most primitive cell and the most advanced animal, the only difference that really matters is the base sequence of the DNA. What is more, the origin of the variations in the DNA that generate those differences is random. Nothing in the behaviour of an animal, nothing of its life experience, alters its DNA in any directed manner such that any acquired characters – strength, knowledge, fears, loves – can be inherited. It must surely press even some biologists' credulity, at times at least, that human beings should have arisen in this manner. But for those who still doubt, the supporting evidence is vast – albeit technical, mathematical and difficult. Unfortunately it could take someone several years to be in a position to understand the subject fully and even begin to make a new contribution.

The behavioural psychologist B. F. Skinner was thus much closer to the truth about the nature of science than Whitehead or Huxley: 'What, after all, have we to show for non-scientific or prescientific good judgement, or common sense, or the insights gained through personal experience? It is science or nothing.'

I would almost contend that if something fits in with common sense it almost certainly isn't science. The reason, again, is that the way in which the universe works is not the way in which common sense works: the two are not congruent. Our brains – and hence our behaviour – have, in evolution, been selected for dealing with the immediate world around us. We are very good at certain types of thinking, particularly that which leads to both simple and quite complex technology and control of our immediate environment. Scientific understanding, however, is not only unnatural: for most of human evolution it was also unnecessary, since, as will be seen (Chapter 2), technology was not dependent on science.

It is precisely the unnatural nature of science that, historically, made it so rare. Unlike science, many features of human behaviour combine unconscious thinking and learning. In marked contrast to their ignorance of physics, most people can carry out the most remarkably complicated actions, such as riding a bicycle – a very difficult problem in Newtonian terms. A remarkable example of how internal mental representations can be used for complex tasks comes from the study of the ways in which Polynesians navigate between distant islands. They use a method involving 'dead reckoning' in which they conceive of the boat as stationary, with the islands moving past it and the stars wheeling overhead. The process has been likened to walking blindfolded between two chairs in a large hall while pointing continually to a third chair off the main path. Such a method of navigation requires no understanding of why it works: it is quite different from one based on science and technology and emphasizes the adaptiveness of human thinking to deal with innumerable problems. While learning is essential, understanding is not.

Unlike science, everyday common-sense thinking is characterized by its naturalness. It involves complex mental processes of which we are usually quite unaware but which allow us to deal with the requirements of daily life. For most of everyday life it works extremely well, but for science it is quite unsatisfactory. It is quite different from scientific thinking, lacking the necessary rigour, consistency and objectivity. Most people regard their ideas about the world as being true without being aware of the grounds for a particular belief. This is quite unlike the self-aware and self-critical methodology of science. Common-sense thinking is also prone to lead to error, particularly when formal problems are posed and when the information available is limited. Indeed, common-sense thinking is not concerned with tackling formal problems or generating general solutions. The differences between common-sense thinking and scientific thinking can be illuminated in two ways: first by looking at the way in which children develop their thinking and then by looking at some aspects of adult thinking.

The perceptual world of the young infant is much more structured than it was previously thought to be. Two-year-olds already understand cause and effect, asking of a broken cup, 'Who broke it?' They also recognize that symbols – words, for example – can stand for things apart from themselves, and they like to put things into categories by colour or size. By their fourth birthday, children appreciate that the appearance of an object – a stone egg for example – may not reveal its true identity. In very general terms, children learn by direct experience, authority, intuition and logic. All of these lead to a common-sense view of the world, but not to a scientific one.

As the Swiss psychologist Jean Piaget has said, every child, at an early stage, fills the world with spontaneous movements and living forces. Waves raise themselves, clouds make wind, and these movements are due to internal and external actions – the objects have a free will of their own. Thus the lake attracts the rivers which wish to go there. Some of the explanations of older children even resemble the physics of Aristotle – for example, the idea that a thrown object is in part moved by the air through which it moves.

There is thus a 'magical' aspect to children's thoughts. In part this may be due to the failure of the infant to distinguish between himself or herself and the world. Whatever the explanation, children believe that mental operations can influence an event that is desired or feared. This is illustrated by the writer and critic Edmund Gosse, who was brought up in a strict Victorian environment in which all imaginative life was forbidden. He was never told stories. He had no friends, and all his reading was pious or scientific. But he wrote that by the age of five or six he had

formed strange superstitions . . . I persuaded myself that, if I could only discover the proper words to say or the proper passes to make, I could induce the gorgeous birds and butterflies in my Father's illustrated manuals to come to life and fly out of the book . . . During morning and evening prayers . . . I fancied that one of my two selves could flit up, and sit clinging to the cornice, and look down on my other self and the rest of us, if only I could find the key.

Piaget has characterized two aspects of children's theory of the world: animism, the tendency to regard objects as living and endowed with will, and artificialism, the idea that everything is made by someone for a special purpose. When a six-year-old is asked what the sun is made of, the reply is 'Of fire.' But how? 'Because there is fire up there.' But where did the fire come from? 'From the sky.' How was the fire made in the sky? 'It was lighted with a match . . .' There is a spontaneous tendency towards animism, for the child to believe as if nature were charged with purpose and as if chance did not exist. When a child says the sun follows us, the child attributes purposiveness to the sun. But when asked 'What is a fork?', the reply is 'It is for eating with' – an artificialist response.

This is evident in relation to the birth of babies. Sometimes the baby is assumed to have existed prior to birth and the child simply asks where it was before. The child may also ask how babies are made, and birth may be conceived by the child as an artificial process of production, like modelling Plasticine, for example. On the other hand, there are often reports of beliefs that babies come from their parents' blood or from the mother's mouth or navel.

One of the most important ideas which lie at the heart of common sense is the idea of cause and effect. Three-year-olds have quite a sophisticated causal understanding of mechanical interactions. The origins of understanding causality have their origins in infancy, and there is now evidence that infants as young as six months perceive causal events. Contrary to David Hume's classical eighteenth-century account, according to which the perception of causality is assumed to be due to the repeated observation of a conjunction between two events, there is evidence that causality is perceived directly almost as a gestalt – that is, as a whole, all at once – in which experience is not important. So, when adults are shown quite abstract stimuli, such as coloured lights with particular movement patterns in relation to each other, causal relations between the lights are proposed even though the observer knows how the stimuli were produced. Thus instead of the appreciation of causality being a result of gradual experience, it seems as if the perceptual system is disposed to assume it. If this were also true for other learning processes, it could require one to abandon much that common sense teaches us.

Children pass through several stages in their competence to perform particular tasks, but they always have satisfactory explanations for their own behaviour. For example, in Piaget's famous conservation task a child sees two identical glass containers filled with water and judges them to contain the same amount of water. As the child watches, one container is emptied into a glass which is taller and thinner, and so the water rises to a higher level. Before children have acquired the concept of conservation of quantity, they will conclude that the amount of water has now increased. Both children who do not understand conservation and those who do will provide what is, from their point of view, a logical explanation for their answer. For example, 'non-conservers' will point out that the water has risen to a higher level in the taller, thinner glass, so clearly there is more water in there. For them their answer is correct and obvious. It is, perhaps, not unlike it being 'obvious' to any reasonable person that the sun moves round the earth.

Older children have quite well-developed ideas about the nature of the world before they are taught science in school. Many of these ideas might be characterized as being naïve or natural thinking, and they are again best illustrated with respect to physics. For example, children suggest that the higher up an object is lifted, the more it weighs, since when it falls to the ground the impact is greater. 'Hot' and 'cold' are considered to be different but related properties: hence some of the cold is thought to leave an ice cube and go into the surrounding water, rather than heat being required to melt the ice and so cool the water. And, to give a biological example, it is widely thought that plants get their food from the soil, rather than from sunlight. (They do, of course, get nitrogen from the soil, but this is not food, for it provides no energy for the life of the plant.) All are common-sense theories, but wrong.

An important feature that has emerged from studies of students' thinking is that inconsistencies in their explanations are usually not noticed, and, if they are noticed, they are not regarded as an important issue. Much of the causal reasoning of students is based on a preference to see change in terms of a simple linear causal sequence or chain of events. This may be the root of the difficulty they have with concepts involving reversibility. They understand how an input of energy can change the state of a substance from a solid to a liquid but not the reverse process, when the liquid solidifies. Studies have shown that a number of key reasoning processes need to be learned before children can grasp the basic nature of the physical world. These include the idea of variables in thinking about causal events, together with the necessity of changing the variables one at a time if a proper comparison of their effects is to be made (in thinking about a simple case of equilibrium, such as in balancing a beam, for example, there are four variables — two weights and two distances from the point of support); the idea of probability and correlation; and the whole idea of abstract models to explain, for example, the solar system or the weather. None of these ideas is really natural, and when children have learned these ideas their success in science tests improves dramatically.

Such studies confirm that scientific thinking differs from everyday thinking not only in the concepts used but in what constitutes a satisfactory explanation: common-sense thinking about motion, for example, is not concerned with the spelling-out in detail of the relationships between terms such as force and velocity — each involving strictly defined and quite difficult concepts — but can be satisfied with vague statements. A further difference is the purpose behind scientific thinking and the thinking of everyday life. In everyday life one is primarily concerned with usefulness, whereas science is concerned with a rather abstract understanding. This is exemplified by Sherlock Holmes when he turns to Watson, who has been castigating him for not knowing about Copernicus and the solar system, and says, 'What the deuce is it to me if you say we go round the sun. If we went round the moon it would not make a pennyworth of difference to me or my work.'

In fact one of the strongest arguments for the distance between common sense and science is that the whole of science is totally irrelevant to most people's day-to-day lives. One can live very well without knowledge of Newtonian mechanics, cell theory and DNA, and other sciences. On the other hand, science can enormously enrich one's life, and in modern society knowledge is essential for innumerable policy decisions that affect our lives (see Chapters 8 and 9).

A formal description of what may be regarded as common sense comes from the American psychologist George Kelly, who has developed what is known as Personal Construct Theory. Central to this theory about the way in which people arrange their knowledge of the world in their everyday life is the idea that they organize information in such a way as to predict future events. Common-sense theories provide mental models of the way in which the everyday world works. People check how much sense they make of the world by seeing how well their model serves them in predicting what will happen. The constructions they place upon events are their working hypotheses which are tested against experience. A person may employ a variety of constructs, some of which may be incompatible with one another, although they are not recognized by that person as being so. Thus at a very low level we may be thought to be doing 'science' in our everyday life by setting up hypotheses and testing them against experience. Cooking is a typical example, since one does experiment; but this is not science since there is no need for theory – only imaginative trial and error is required to achieve the right 'taste'. Doing science, on the other hand, requires one to remove oneself from one's personal experience and to try to understand phenomena not directly affecting one's day-to-day life, one's personal constructs. In everyday life, one requires no construct as to why bodies fall when dropped or why children may or may not resemble their parents; it is sufficient that they do so. Common sense provides no more than some of the raw material required for scientific thinking.

At its simplest most human actions involve forming a goal and modifying one's actions in order to achieve the goal. The value of this simplified model is that it emphasizes the common-sense nature of our behaviour and what we were designed for. The model requires no science as such, and that is why early technology could be so successful. Another feature of this scheme is that precision, accuracy and completeness of knowledge are seldom required – quite unlike science. We make decisions based upon what is in our memory – a memory that is, as will be seen, biased toward overgeneralization of the commonplace and overemphasis on the discrepant or rare cases.

Whereas scientific theories may be judged in terms of their scope, parsimony – the fewer assumptions and laws the better – clarity, logical consistency, precision, testability, empirical support and fruitfulness, lay theories are concerned with only a few of these criteria and are seldom explicit or formal, or consistent, and are often ambiguous. The explicit or formal nature of scientific theories is not only important in its own right but points to a crucial feature of the scientific process: the self-aware nature of the endeavour. This self-aware aspect of doing science, as distinct from other activities, makes science different from common sense almost by definition, since, again almost by definition, common sense is unconscious. The scientist is always aware of 'doing science', and with that self-awareness go a number of assumptions which are seldom made explicit. They include some of the characteristics of science listed above but also include ideas that put a high value on elegance and generality (Chapter 6).

Objectivity as distinct from subjectivity is a conventional means of characterizing scientific thinking. It is important – indeed essential – to try to separate evidence from theory and also to be able to look objectively at a theory, to recognize it as something on its own. But the idea of scientific objectivity has only limited value, for the way in which scientific ideas are generated can be highly subjective, and scientists will defend their views vigorously. Being objective is crucial in science when it comes to judging whether subjective views are correct or not. One has to be prepared to change one's views in the face of evidence, objective information. It is, however, an illusion to think that scientists are unemotional in their attachment to their scientific views (Chapter 5): they may fail to give them up even in the face of evidence against them. Another crucial difference from common-sense or lay theories is that scientific theories involve a continual interplay with other scientists and previously acquired knowledge for scientific ideas are directed not just at a particular phenomenon in everyday life but at finding a common explanation for all the relevant phenomena, and an explanation which other scientists would accept.

Associated with lay theories is a tendency to adapt and modify the theory too hastily in relation to the way people live, because people want to believe in a just and more or less ordered world over which they have some control. Many conclusions are influenced by the emotional content of the data. Bertrand Russell proposed that 'popular induction depends upon the emotional interest of the instances, not upon their number.' Examples of this abound in everyday life. Suppose that, via consumer reports and your local and trusted garage, you have carefully researched what car to buy and have settled on model X. And then you meet a close colleague and tell him of your decision. If he then reacts with shock and relates his own terrible experience with car X, listing all the problems he had, would you really be unaffected? Even though his account is but one in a large number, you will have great difficulty ignoring his advice.

Research into how people reason about complex issues of genuine importance such as crime and unemployment again emphasizes the difference between common-sense thinking and more formal scientific thinking. At the extremes there are two very different attitudes towards knowledge. One pole is the comfortable ignorance of never having considered that things could be otherwise; the other is a continual self-aware evaluation of the evidence and subsequent modification of views. These reflect the distinction between knowing something to be true and contemplating whether one believes it to be true or not. Only a minority (about 15 per cent) appear to have the latter capacity but scientists – even though they may not like to – have to adopt this approach.

The processes by which we make deductions in everyday life, such as about the cause of a particular event, are often carried out by processes of which we are unaware. Such processes are poorly understood, and it is notoriously difficult to mimic 'common sense' on a computer. For example, if you leave your house one morning and notice that the grass is wet, you are almost sure it rained during the night. But if you then learn that the sprinkler was left on all night, your confidence in the 'rain hypothesis' is greatly diminished. It is hard to program this into a computer. The psychologist Johnson-Laird claims that common-sense thinking is based neither on formal logical rules for inference nor on rules that contain specific knowledge. It seems that the way we reach valid conclusions from a set of premises is to construct mental models. The mind then can manipulate the model it has produced and try out various alternatives. Conclusions can be drawn from the model which can then be tested. Consider the following problem, which is hard to solve by common-sense thinking. In a room of archaeologists, biologists and chess-players, if none of the archaeologists is a biologist and all the biologists are chess-players, what inferences can be drawn? One can try various models to see which inference can be made, rather than proceed by formal logic. The only correct inference is that 'Some chess-players are not archaeologists.' This case shows how difficult formal reasoning can be.

We may like to see ourselves as naturally rational and logical, but there is a lot of good evidence that this is not always so. While in everyday thinking the mind can show some adherence to logical rules, these can be influenced by the nature of the problem, and so the formal rules break down. This can be illustrated by what is now recognized as a classic and seminal experiment. Imagine you are presented with four cards, each of which has a letter on one side and a number on the other. The four cards when placed on the table show A, J, 2 and 7. Your task is to decide which cards should be turned over in order to determine the truth or falsity of the following statement: 'If there is a vowel on one side of the card then there is an even number on the other side.' Most people correctly turn over the card bearing the A, and some turn the card with 2 on it. Few choose the card with 7, even though this is a logical choice – for if there were a vowel on the other side of the 7 the rule would be falsified. Turning over the J or the 2 tells one nothing. Whatever is on the other side of the 2 will not provide useful information, since whether or not it is a vowel or a consonant will not determine the validity of the rule. This experiment shows in addition the preference that people – including scientists – have for trying to confirm hypotheses, rather than for trying to refute them.

One area of day-to-day thinking which has been shown to be particularly prone to errors is that which involves probabilities and judgements which have to be made on the basis of uncertain information. Many scientific investigations have to be done under precisely such conditions, and the scientist has somehow to become free from the all-too-common errors.

Children have a limited understanding of chance: they believe that outcomes of games based on chance can be influenced by practice, intelligence and effort. Adults, too, have difficulty with probabilities and the nature of chance. If you are playing roulette and red has come up five times running, is the chance of black greater on the next spin? The answer is 'no', and the contrary expectation is known as the 'gambler's fallacy'. Again, if, in spinning a coin, heads has come down ten times running, the probability of a tail or a head at the next spin is still 0.5 – evens. The coin has no memory. Given an evenly balanced coin, many people believe that a sequence H-T-H-T-H-T is much more likely than H-H-H-H-H-H, whereas in fact both are equally likely.

Correct probability judgements are often counter-intuitive. Striking coincidences often lead to ideas of supernatural forces at play. For example, to hear that a woman had won the New Jersey lottery twice in four months seemed remarkable, and the odds against it were claimed to be 17 trillion (17×10^{12}) to 1. But further analysis showed that the chance that such an event could happen to someone, somewhere, in the United States was about one in thirty, because so many people take lottery tickets. Another example is that it only requires twenty-three people to be together in a room for the probability of two of them having the same birthday to be one in two.

There was, a little while ago, a spate of articles in news-papers in the USA which suggested a link between teenage suicide and a game called 'Dungeons & Dragons'. It was said that the game could become an obsession and lead to a loss of a sense of reality. Evidence to support this claim was that twenty-eight teenagers who often played the game had committed suicide. However, the game had sold millions of copies, and probably as many as 3 million teenagers played it. Since the annual suicide rate for teenagers is about twelve per 100,000, the number of expected suicides in a teenage population of 3 million is about 360. So, finding twenty-eight such suicides has little or no significance on its own.

These examples of failure to appreciate the nature of probabilities and statistical thinking are particularly important when it comes to assessment of risk. It is, for example, rarely appreciated that it is almost impossible to ensure that a drug does not cause a death rate of, say, one in 100,000. Indeed the basis for clinical trials is rarely appreciated. In order to show the efficacy of a particular drug or medical treatment, it is essential to follow a rigorous procedure for the selection of a sample group, some of whom will be treated and some of whom will not. The assignment to the treated or non-treated group must be random, and wherever possible doctors themselves should not be aware of who is being given which treatment. Moreover, the results will require a careful statistical analysis. Such expensive trials are essential, but a 1 in 100,000 death rate due to the drug would require an enormous sample. Anecdotal collections of cases in which cures of, for example, cancer, are claimed can be very misleading.

An important class of error is based on what is known as representativeness – that is, the degree to which one event is representative of another is judged by how closely they resemble one another. For example, experimental subjects were given descriptions of men taken from a group that comprised 70 per cent lawyers and 30 per cent engineers and were asked to assess the profession of each man described. Even though the subjects knew the composition of the group, and thus should have seen that the probability of being a lawyer was more than twice that of being an engineer, the subjects nevertheless consistently judged a description to refer to an engineer if it contained even the slightest hint, no matter how unconvincing, of something that fitted their stereotyped image of an engineer. They ignored the probabilities involved in selecting a single case from a population of known composition. And this tendency was even more pronounced when assessing the reliability of small samples. Subjects are, for example, very bad at judging the likelihood that the number of boys being born each day would be greater than 60 per cent in a large and a small maternity hospital. They usually thought that there would be no difference, whereas in fact, with a small sample, the changes in the percentage of boys at a small hospital are very much greater, because each birth represents a greater percentage of the total.

In fact most of us have poor intuitive understanding of the importance of chance where small numbers are involved.

Representativeness also results in people having much greater confidence in their ability to predict than is in fact warranted. A superficial match between, for example, the input and the outcome generates a confidence which ignores all those factors which would limit the validity of the prediction. For example, staff at medical schools select students and believe in their ability to select correctly. But they can later judge only those students whom they have selected: they cannot compare them with those whom they rejected. This is well illustrated by psychologists' confidence in their own ability to select the best candidates at interview even though they know of the extensive literature showing quite conclusively how unreliable the interviews are. They cannot restrain their own convictions about their own reliability.

Another example is where people judge frequency according to a method which depends on the information available to them – that is to say, they estimate frequency in terms of the examples that come to mind. Thus most people believe that there are more words beginning with the letter R than there are words which have R as the third letter, because words beginning with R are easier to think of. Similarly they give a much lower estimate for $1 \times 2 \times 3 \times 4 \times 5 \times 6 \times 7 \times 8$ than for $8 \times 7 \times 6 \times 5 \times 4 \times 3 \times 2 \times 1$, and in both cases it is far too low. Typical answers are around 500, whereas the correct answer is 40,320. The plausibility of the scenarios that come to mind serve as an indication of the likelihood of an event. If no reasonable scenario comes to mind, the event is deemed impossible or highly unlikely; if, however, many scenarios come to mind, the event in question appears probable. Even physicians tend to have distorted ideas about the dangers of various diseases that are frequently referred to in medical journals, irrespective of their true incidence

We tend to generalize from our own experience, and so there is a tendency to believe illusory correlations ranging from 'fat people are jolly' to 'if you wash your car it will rain soon afterwards' and all sorts of theories about illness. Even psychologists have been known to find correlations between projective tests when none were later shown to exist.

However, simple associations are probably very useful in everyday life.

There is in general a preference for simple rather than complex explanations. It is possible to understand such a predisposition in evolutionary terms. For primitive humans it would have been an evolutionary advantage to learn about the environment rapidly and to infer causal relationships. Selection for a brain that could directly appreciate probabilistic events and counter-intuitive results would seem to be extremely unlikely in a hostile environment where rapid and immediate judgements are required. And the use of tools and the development of technologies such as metalworking and agriculture do not require scientific thinking. But to do science it is necessary to be rigorous and to break out of many of the modes of thought imposed by the natural thinking associated with 'common sense'.

2
Technology is not Science

Much of modern technology is based on science, but this recent association obscures crucial differences and the failure to distinguish between science and technology has played a major role in obscuring the nature of science. To put it briefly, science produces ideas whereas technology results in the production of usable objects. Technology – by which I mean the practical arts – is very much older than science. Unaided by science, technology gave rise to the crafts of primitive man, such as agriculture and metalworking, the Chinese triumphs of engineering, Renaissance cathedrals, and even the steam engine. Not until the nineteenth century did science have an impact on technology. In human evolution the ability to make tools, and so control the environment, was a great advantage, but the ability to do science was almost entirely irrelevant.

For some historians, science began whenever and wherever humans tried to solve the innumerable problems of dealing with the environment. For them, technology, starting with toolmaking, is problem-solving and hence science. In fact the crafts associated with agriculture, animal domestication, metalworking, dyeing and glass-making were present thousands of years before the appearance of what we think of as science. In *The Savage Mind*, the French anthropologist Claude Lévi-Strauss argues that 'Each of these techniques assumes centuries of active and methodical observation, of bold hypothesis, tested by means of endlessly repeated experiments.' Put in this way, he makes it sound like a formula for doing science and makes it seem that primitive technology involved mental processes very similar to those of science. But did this early development of technology involve bold hypotheses?

Lévi-Strauss has no doubt that neolithic or early historical man was heir to a long scientific tradition. If he is right, then there is a paradox, as he forcefully points out. If neolithic culture was inspired by 'scientific' thinking similar to our own, it is impossible to understand how several thousand years of stagnation intervened between the neolithic revolution and modern science. For Lévi-Strauss there is only one solution to the paradox, namely that there are two distinct modes of scientific thought, two strategic levels at which nature is accessible to scientific enquiry: one roughly adapted to perception and imagination; the other at a remove from it. The 'science of the concrete . . . was not less scientific and its results no less genuine. They were secured ten thousand years earlier and still remain at the basis of our own civilization.' But, as I will try to show, Lévi-Strauss's two modes of thought are, in fact, science and technology – and technology requires no understanding or theory of the kinds provided by science.

Agriculture was already in progress at about 7000 BC when man passed from hunting and gathering to food-producing. Cattle were probably domesticated at this time, but there is no reason to believe that the farmers had any more understanding of the science involved in agriculture than most Third World farmers have today. They relied on their experience and learned from their mistakes. Of course there was inventiveness, but this inventiveness was of the same kind involved in primitive toolmaking: it was an acquired skill based on learning and is closely linked to common sense. There is no reason to distinguish such inventiveness from an extension of the ability of chimpanzees to manipulate their environment to achieve a particular goal. Classic examples of such behaviour include their ability to join two sticks together to get bananas from a hook too high up for them to reach with their hands. This in no way lessens the achievements of early technology, but it does help distinguish it from science.

By 3500 BC there was already a high degree of competence in metalworking, and by 3000 BC Mesopotamian craftsmen mixed copper and tin in varying proportions to produce different sorts of bronze. The kilns must have produced temperatures of over 1,000°C. In the case of glassworking, there is a text from round 1600 BC found near Baghdad which gives a description of how to make a green glaze. Essentially it is a recipe. It begins, 'Take a mina of zuku glass together with ten shekels of lead, fifteen shekels of copper . . . ,' and it continues with detailed instructions as how to proceed: 'Dip the pot in this glaze, then lift it out, fire it and leave it to cool. Inspect the result: if the glaze resembles marble, all is well. Put it back in the kiln again . . .' Mixed in with such practical injunctions there were also ritual 'magical' actions. For example, from the seventh century BC there are instructions that the glass-furnace must be built at an auspicious time, a shrine must be installed and the deities placated. 'When laying out the ground-plan for the glass-furnace, find out a favourable day in a lucky month for such work . . . Do not allow any stranger to enter the building . . . Offer the due libations to the gods daily.'

Copper-making was well developed on the coast of Peru as early as 500 BC, many hundreds of years before the arrival of the Spaniards. Evidence from furnaces from around AD 1000 suggest that smelting was associated with solemn rituals and offerings to deities.

The technological achievement of the ancient cultures was enormous, and Lévi-Strauss is right to pose the question of how it was achieved. But whatever process was involved, it was not based on science. There is no evidence of any theorizing about the processes involved in the technology nor about the reasons why it worked: for example, it was enough to know that adding charcoal to the molten mixture would accelerate the smelting of iron. Metalworking was an essentially practical craft based on common sense. The goals of the ordinary person in those times were practical ends such as sowing and hunting, and that practical orientation does not serve pure knowledge. Our brains have been selected to help us survive in a complex environment; the generation of scientific ideas plays no role in this process.

As technology became more advanced and resulted in more complicated inventions like the telescope, compass and steam engine, it might be thought that science, which was by then itself quite advanced, would have made significant contributions to these inventions, even if it played no role in early primitive technology. This is not the case. As will now be shown, science did almost nothing to aid technology until the nineteenth century, when it had an impact on synthetic-dye production and electrical power.

Galileo understood quite clearly that the technology of his time, the early seventeenth century, was not based on science. The inventor of eyeglasses and the telescope is unknown, and Galileo comments on this: 'We are certain the first inventor of the telescope was a simple spectacle-maker who, handling by chance different forms of glasses, looked, also by chance, through two of them, one convex, one concave, held at different distances from the eye; saw and noted the unexpected result; and thus found the instrument.' Galileo himself improved the telescope by trial and error, aided by his skill as an instrument-maker, and not by his understanding of optics.

Francis Bacon, unlike his contemporary Galileo, was confused about the relation between science and technology and he drew no real distinction between them. 'Science also must be known by works . . . The improvement of man's mind and the improvement of his lot are one and the same thing.' Science and technology are here conflated. (Compare this with Archimedes' contempt for the practical, described in the next chapter.) The three inventions which he identified as the source of great changes in Renaissance Europe – printing, gunpowder and the magnetic compass – were Chinese imports and owed nothing to science; nevertheless, he believed that scientific accomplishments would transform human activity through technological change, though he did not have a single example to support his case.

The history of technology is largely an anonymous one, with few honoured names – again, unlike science. Neither learning nor literacy was relevant. Who, for example, was the unknown genius who realized that a thin piece of metal coiled into a spiral could be made to drive a machine as it unwound? Spring-driven clocks were being made early in the fifteenth century. Other crucial inventions were machines for cutting the teeth in wheels to make gears. Both the screw and the gear were invented by the Greeks – Archimedes had used a spiral screw for raising water in the third century BC – but the ability to make both reliably, in metal, required the construction of special and ingenious machines that in the fifteenth century gave rise to the metalworking lathe.

The wheel also illustrates a nice absence of relation between technology and science, for why does a wheel make it easier to move a load? The answer is moderately subtle: the wheel reduces the friction between the object moved and the ground. Most of the work required to move an object over a surface is needed to overcome friction between the object and the surface. By using a wheel, the friction is reduced both by having an axle which is smooth and so reduces friction and by introducing a rolling motion at the surface. But that understanding, based on science, is completely unnecessary for either the invention of the wheel or the appreciation of its usefulness.

The mechanics of building again illustrates the independence of technology and science until recent times. Statics, the science of the forces acting on a body at rest, like an arch or a bridge, was founded by Archimedes when he devised formulae for the equilibrium of simple levers and for determining the centres of gravity of simple objects. It was not until some 1,800 years later that further progress was made, by the Dutch mathematician Simon Steven, who, in the sixteenth century, showed how to analyse more complex combinations of forces. How to calculate correctly the forces acting on a structure became clear in the eighteenth and nineteenth centuries, but this knowledge began to be applied to building structures only in the nineteenth century: none of the buildings constructed before that time made use of any scientific principles that are used in modern engineering. They probably did make use of what may be thought of as 'The Five Minutes Theorem': if a structure was built and remained standing for five minutes after the supports had been removed, it was assumed it would stand up forever.

All the beautiful cathedrals with their great domes and high naves were built by engineers who based their buildings on practical experience, not on science. The early iron bridges were also constructed on a purely empirical basis. So the bridge designed in the 1850s by Robert Stephenson and William Fairbairn to span the Menai Strait in North Wales – the first box-girder bridge – was based on experiments. A series of models was used to establish the design. The theory which could have provided an analytical approach to designing the structure had been published a few years earlier, but it was ignored. Technology may well have used a series of ad hoc hypotheses and conjectures, but these were entirely directed to practical ends and not to understanding. There was no attempt at generality

Science by contrast has always been heavily dependent on the available technology, both for ideas and for apparatus. Technology has had a profound influence on science, whereas the converse has seldom been the case until quite recently.

The invention of the steam engine, pendulum clock and navigational techniques requires special examination, since science did play a role here, but not necessarily in terms of understanding. However, the very rarity of such special cases underlines and strengthens the main thesis.

The origin of the steam engine can be thought of as owing more to the blacksmith's world than to the Royal Society and its scientists. James Watts's steam engine of 1775 was a major modification of the Newcomen engine (1712) which had been in wide use for sixty years. Newcomen's engine was based on the condensation of steam in a cylinder – this caused a partial vacuum, and atmospheric pressure then forced a piston down into the cylinder. Thus the working stroke involved the piston moving into the cylinder, whereas in Watts's steam engine the pressure of the steam drove the piston outwards. Scientists had for centuries been fascinated by the very idea of a vacuum. In the 1690s Denis Papin, a French scientist, had devised a machine for making a vacuum in a cylinder containing a piston, based on the condensation of steam. He realized that the condensation process could be used to provide useful work. What is unclear is whether or not Newcomen, an ironmonger, was aware of Papin's work and current ideas about atmospheric pressure. Even if he was, his engine was very different from Papin's simple cylinder and piston. More important, his engine was not based on any theoretical consideration; rather, the apparatus used for a scientific experiment may have formed the basis for a technological invention.

Another area where it might reasonably be thought that science did have an impact on technology is timekeeping and navigation. Galileo introduced the pendulum to clocks. The story goes that, at the age of nineteen, he had noticed that a swinging altar lamp always took the same time to move from one side to the other, no matter whether the swing was large or small. He did not have to understand why this was so in order to recognize its value in timekeeping, but this does come very close to science affecting technology. Similarly, I have to recognize that, before there was an accurate clock that could be carried on board a ship, navigators needed some training in mathematics: to find longitude at sea required precise observations on the moon and quite subtle calculations.

The motivations behind technology and science are very different. The final product of science is an idea, or information, probably in a scientific paper; the final product of technology is an artefact – the clock or the electric motor, say. Unlike science, the product of technology is measured not against nature but in terms of its novelty and the value that a particular culture puts on it. Whether or not it is true, statements such as that of Karl Marx to the effect that inventions since 1830 could be thought of as being 'for the sole purpose of supplying capital with weapons against the revolts of the working class' could not conceivably be made about scientific ideas.

A more interesting general question is: what drives technological and scientific advance? For technology it is the demands of the market-place or advancing technology 'making' the need. Inventive activity is, it seems, governed by the expected value of the invention – inventions peak when investments peak – and patents also illustrate a clear difference between science and technology, for one cannot patent scientific discoveries or ideas. Oliver Lodge disliked the idea of patenting his ideas on radio waves, as patenting is the antithesis of the openness which scientists want. The reward for the inventor is money; for the scientist it is esteem.

In earlier times, the ethos of the craftsmen was like that of a guild: learning was by apprenticeship, outsiders were excluded and secrecy was essential. In this, too, it differed from science, for which openness, controversy and public access to knowledge are characteristic features. Yet another difference lies in the selection criteria that determine success: for technology, success is related to wants and needs; for science, success depends on correspondence with reality.

Technology has its own evolutionary history. The historian of technology George Basalla has adopted a biological approach to technology, reviewing its history in evolutionary terms. An artefact is regarded as the fundamental unit, and continuity prevails – different versions result from modification of the original object. By contrast, ideas, not artefacts, are the fundamental units in science. A key feature in the evolution of technology is diversity, which is conventionally ascribed to necessity and utility. But the variety is astonishing, and even Marx was surprised to learn that 500 different kinds of hammer were produced in Birmingham in 1867. Was this diversity really necessary and useful? In general terms, Basalla argues that technology does not always exist primarily to supply humanity with its needs; rather, the need often develops only after the invention. For example, the invention of the internal-combustion engine gave rise to the necessity for motor transportation.

The story of the wheel illustrates his point. Only some thousands of years old (compared to the one and a half million for the making of fire), the wheel probably developed from the rollers that were used to move heavy objects. Evidence that wagons were used for transport dates from round 2000 BC, about one thousand years after the wheel's first appearance in Europe and Asia. In the Americas and southern Africa, for example, the wheel did not appear until modern times. The puzzle is Central America: wheeled transport arrived only with the Spaniards, in the sixteenth century, but long before, from the fourth to the fifteenth centuries, small figurative sculptures were fitted with axles and wheels to make them mobile. An explanation as to why this invention of the wheel was not developed for transport is that there was no need since, except in Peru, there were no roads, and there were also no large domesticated animals to pull heavy loads. Again, between the third and seventh centuries the camel performed the role of wheeled vehicles in the Near East and North Africa. The wheel is not a universal need.

The interaction between science and technology in recent times has been illuminated by Basalla's discussion of the history of radio communication. Electromagnetic waves had their origin not in experiment but in the equations which James Clerk Maxwell developed in the second half of the nineteenth century. His equations initially dealt with all that was known about electricity and magnetism, but for mathematical consistency he introduced a new term that effectively implied the propagation, with the speed of light, of electromagnetic waves. He made no effort to verify the existence of such waves, however. His theory essentially put Michael Faraday's ideas about electricity and magnetism in a mathematical form, and at the same time provided a completely new conception of electromagnetism by considering how Faraday's lines of force were produced and what medium they required for their propagation. In spite of the highly mathematical nature of his analysis, he presented the theory in terms of physical models that related to the technology of the time – so much so that the French mathematician Henri Poincaré remarked that 'one seemed to be reading the description of a workshop with gearing, with rods transmitting motion and bending under the effort, with wheels, belts and governors'. It is ironic that Maxwell's new ideas were visualized in terms of the rather old-fashioned technology of his age.

Heinrich Hertz's contribution, in 1888, was to demonstrate the propagation of electromagnetic waves. Yet is was not Hertz but Oliver Lodge, who was doing similar experiments, who recognized their importance for telegraphy. His interest was rather reluctant, and it was left to Marconi to pursue the commercial exploitation of Hertzian waves. Just before Marconi's invention, the English scientist Karl Pearson had, in 1892, written in his book *The Logic of Science* that he regarded electromagnetic waves as having no useful application.

The very natures of scientific and technological thinking

are dissimilar. Many aspects of technology are visual and non-verbal, which is quite unlike scientific thinking. It is not that scientists do not visualize structures, concepts and mechanisms, but exposition is fundamental to science and the images must be translated into language and symbols, particularly mathematics. Unencumbered by verbalized theories, the designers of technology bring together, in their minds, different elements in new combinations. In contrast to science, technological knowledge from the Renaissance until the nineteenth century was carried in books which were dominated by illustrations – the information was largely carried in pictorial form. Many of the books carried numerous illustrations of mechanical linkages, assemblies of gears and cams, and machines themselves, such as pumps. Curiously, there were claims that all these mechanical arts rested on the firm foundation of mathematics, but quite the contrary was true: there is no evidence for the use of either geometry or arithmetic in the design of the machines. It seems that it was true even then, as now, that claims that designs were based on science gave them greater respectability. Visual thinking also dominated industrial design. Science offered no guidance to the early designers of motorcycles, for example – it could not tell them where to put the engine, battery and fuel tank in relation to one another.

Engineering, even today, should not just be construed as merely applied science. The relationship between science, technology and industrial success in modern societies is complex. Many have puzzled as to why Japanese industry should have been so successful. It has been suggested that its success is based not on science but on its ability to apply science. The transistor, invented in the United States and the basis of modern electronics, was initially perceived as a replacement of the old thermionic valve; the idea of an integrated circuit developed only slowly. There is no doubt that the invention of the transistor depended on science, but its exploitation was rather different: the Japanese showed that a strong scientific base was not necessary for a successful manufacturing industry.

3
Thales's Leap: West and East

The peculiar nature of science is responsible for the fact that, unlike technology or religion, science originated only once in history, in Greece. Most scholars are agreed that science had its origin in Greece, though those that equate science with technology would argue differently. This unique origin is important for understanding the nature of science, since it makes science quite different from so many other human activities, for no other society independently developed a scientific mode of thought, and all later developments in science can be traced back to the Greeks. It is my intention not to try to account for this single origin but to emphasize how rare science is in human cultural history and also to use its origin to illuminate some of the special characteristics of scientific inquiry.

Thales of Miletos, who lived in about 600 BC, was the first we know of who tried to explain the world not in terms of myths but in more concrete terms, terms that might be subject to verification. What, he wondered, might the world be made of? His unexpected answer was: water. Water could clearly change its form from solid to liquid to gas and back again; clouds and rivers were in essence watery; and water was essential for life. His suggestion was fantastical perhaps, but such unnatural thoughts – contrary to common sense – are often the essence of science. But more important than his answer was his explicit attempt to find a fundamental unity in nature. It expressed the belief that, underlying all the varied forms and substances in the world, a unifying principle could be found. The possibility of objective and critical thinking about nature had begun. Never before had someone put forward general ideas about the nature of the world that might be universal, ideas that tried to explain the nature of the world in a way quite unlike the explanations provided by all-pervasive myths. For the first time there was a conviction that there were laws controlling nature, and that these laws were discoverable. Together with an emphasis on rationality, such ideas were to be crucial to the success of science and its survival later in the West. This was one of the most exciting and important ideas in the entire history of mankind. But, even more important, this idea was open for discussion and debate. It was a wonderful leap that was to free thinking from the strait-jacket of mythology and the grip of relating everything to man. Here, too, for the first time, attention was focused on the nature of the world with no immediate relevance to humankind. Human curiosity had hitherto been entirely devoted to man's relation to nature, and not to nature itself. It is with the Greeks that man and nature are for the first time no longer perceived as inextricably linked and there begins a distanced curiosity about the world itself.

While giving the honour of being effectively the first scientist to Thales of Miletos, one recognizes that Thales was himself a philosopher and heir to an intellectual tradition whose origins are obscure. He cannot have been totally unaware of the achievements of the Egyptians and particularly the Babylonians with respect to the use of mathematics. Miletos, where Thales lived, was the main harbour and the richest market of Ionia, trading with Phoenicia, Egypt and many other countries. This would have provided a rich and varied environment. In addition, the Ionians were colonists and may perhaps be assumed to have the intellectual vigour and the freedom from well-established ideas that characterize many immigrant communities. The Greeks, unlike the Jews, had no dogmas like the Old Testament to constrain their thinking, though they did have plenty of myths.

It was also Thales who established mathematics as a science, irrespective of how much he might have learned from the Babylonians and Egyptians, who had established arithmetic procedures and the elements of geometry for their practical needs. The Babylonians knew elements of geometry as early as 1700 BC, and had tables listing the sides of right-angled triangles – they thus must have been aware of the key features of Pythagoras's theorem which states that the square of the hypotenuse is the sum of the squares of the other two sides. Egypt contributed little to the advancement of mathematics, but used it for practical problems of measurement. Thales, by contrast, turned these tools of measurement into a science. He put forward a number of basic propositions: that a circle is bisected by its diameter; that, if two straight lines cut each other, the opposite angles are equal; and that the angle inscribed in a semicircle is a right angle. Here, for the first time, were general statements about lines and circles – statements of a kind never made before. They were general statements that applied to *all* circles and lines everywhere, and that is the generality to which science aspires. The Greeks transformed a varied collection of empirical rules for calculation into an ordered abstract system. Mathematics was no longer merely a tool used for practical problems: it became a science.

Thales's contemporary in Miletos, Anaximander, did not find Thales's ideas about water persuasive. To Anaximander it seemed that air was a much better candidate for being the primary substance of which all things were made. And so began the sort of claim and counterclaim for the understanding of nature which eventually gave rise to modern science. There was, even so, a crucial ingredient still lacking: experimental method.

With Thales and the later Greeks there came the transition from explanations by means of myths to explanations which were self-consistent and open to critical analysis. This constituted a very big change. While myths do provide explanations to questions about 'how' and 'why', they are defective from at least two points of view: the problem being addressed may not be explicit, and the proposed solution may rest on arbitrary assumptions whose applicability is not specified. For example, the circumstances under which the Babylonians believed that Marduk split the primeval water goddess Tiamat to make the sky and its celestial waters on one side and the 'great abode' on the other are not made explicit. Similarly, the Egyptian explanation that the movement of the sun is due to the god Ra rowing a boat across the sky is a story, not an explanation in scientific terms: it is neither verifiable nor falsifiable. By contrast, Aristotle's discussions about the shape and position of the earth and its movement, even though they were wrong, belonged to a quite different class. Together with these new kinds of explanation came a critical appreciation of the nature of explanation itself, and the requirement for logical consistency. It was no longer acceptable to suggest that the earth does not move because it is supported by, say, air or water; for what, in turn, supports that?

The stage for science had been set, and for the first time there were named actors, with strong views and personalities. This was a radical departure, for Egyptian and Babylonian medicine, mathematics and astrology can, apparently, be combed in vain for examples of a text where an individual author explicitly distances himself from and criticizes the received tradition in order to claim originality for himself; whereas in Greece this became a normal procedure. Perhaps this arose from the similar requirement for recognition by the Greek poets, but, whether this is so or not, scientists and philosophers typically appear in the first person. This may also be related to the fact that many Greek citizens acquired experience in the evaluation of evidence and argument in the context of politics and law. So a critical tradition of crucial importance was established and, one after the other, pre-Socratic philosophers implied that no one else had got the answers right. Authority was challenged, and the ideas of individuals about the nature of the world became dominant. The admiration of one's peers is one of the major rewards in science, and this became possible only when science became the work of individuals who adopted the crucial first-person singular.

Aristotle's science, which became dominant, can make difficult reading. For example, he distinguishes four kinds of causes, only two of which relate easily to a modern reader: cause in the sense of one thing's influence on another, and cause in the sense of the function that something serves. Nevertheless, his science accords with a reasonably common-sense picture of the world. He consciously applied a maxim that in the search for explanations it is necessary to start from what is familiar and that deductions in science can proceed from principles intelligible in themselves. One should view Aristotle's situation with sympathy, for how was he to know that the world is constituted in a way that bears no relation to common sense? Aristotle's world is made up of four basic elements – earth, fire, air and water – and each of them has two of the four primary qualities – wetness, dryness, coldness and hotness. All these are drawn from everyday experience. Movement of objects now finds a natural explanation. Fire moves upwards and earth downwards to their natural places. The earth is at the centre of the universe, and the heavenly bodies are embedded in a series of concentric spheres around it. Circular motion is regarded as perfect, and this describes the movement of the sun and the heavens. Aristotle's contribution to biology was to open up many areas – comparative anatomy, embryology and animal behaviour – and to make an enormous number of observations. His teleological explanations also made sense, since they implied that natural phenomena had an end in view. Why do ducks have webbed feet? In order to swim. Aristotle never arrived at the fundamental requirement of doing experiments in relation to theories; however, he came close by providing the basis for thought experiments, such as thinking about what direction the earth would move in if the heavens stood still.

Aristotle also clearly recognized one of the key features of early science: it offered no reward other than intellectual gratification. 'Thus, since men turned to philosophy in order to escape from a state of ignorance, their aim was evidently knowledge, rather than any sort of practical gain. The evidence of history confirms this: for, when the necessities of life were mostly provided, men turned their minds to this study as a leisure-time reaction.'

Most of Greek science turned out to be wrong – misconceptions about motion, embryological development and the place of the earth in the heavens. That is no disgrace, for being wrong is a constant feature of scientific method. However, at least two giants stand out. Their achievements are almost as great as Thales's great leap. Euclid's geometry and Archimedes's mechanics were fundamental to further scientific advance, and one may speculate, with some concern, how the scientists of the Renaissance would have fared without them.

Euclid, who lived around 300 BC, was not the inventor of geometry, for many propositions had been known for a long time before him. His achievement was to follow through Aristotle's demand for a logically derived science based on a minimum number of postulates, which had to be taken as given; his five postulates are undemonstrable but taken to be true. Most of Euclid's postulates evoke little surprise and seem quite sensible – for example, that all right angles are equal, and that a circle can be constructed when its centre and a point on it are given. However, the fifth postulate is rather different: 'If a straight line falling on two straight lines makes the interior angles on the same side less than two right angles, the two straight lines, if produced indefinitely, meet on that side on which are the angles less than the two right angles.'

Another way of stating this postulate is to say that parallel lines never meet, and either formulation may seem to be obvious. But what is not at all obvious is that, given the other postulates, it cannot be proved. Euclid's genius recognized that proof was impossible and that this needed to be included among the postulates. Of course, another postulate could be formulated to replace it, but some equivalent postulate *is* necessary. Now, given the five postulates, the richness of Euclidean geometry could be deduced. Also, a model for a hypothetico-deductive science as proposed by Aristotle was established; that is, given a number of laws and basic assumptions, a large and varied number of conclusions could be drawn. (It is worth noting that, given Euclid's postulates, common sense would not alone be sufficient to derive the theorems of geometry.)

Archimedes studied under the disciples of Euclid in Alexandria in the third century BC and was the first applied mathematician: he applied mathematics to understanding how the world works. He laid the foundations not only for statics – that is, the study of non-moving forces in equilibrium, such as the forces exerted by levers and weights, which is the basis of all structural engineering – but also for hydraulics – the study of forces acting on bodies in water. Archimedes invented machines such as the compound pulley and a hydraulic screw for raising water, but he himself, in the Greek tradition, did not value such achievements: according to Plutarch, he regarded as 'ignoble and sordid the business of mechanics and every sort of art which is directed to use and profit; he placed his whole ambition in those speculations the beauty and subtlety of which are untainted by any admixture of the common needs of life.' This is an early and crucial example of the differing attitudes to pure and applied science. Apparent uselessness is one of early science's pecularities, for what use was it to Thales that all the world was made of water, or to Archimedes that he understood why some bodies floated, or to Aristotle that the heart was the first organ in the embryo to develop?

While Archimedes made important contributions to mathematics – he defined the Archimedean spiral and made a good approximation for π, the ratio of the circumference to the diameter of a circle – it is his mechanics and hydrostatics that are so impressive. In these he achieved for the first time, for physics, what Euclid had done for geometry. He, like Euclid, begins with definitions and postulates and then proves certain propositions. This approach was applied to mechanics, where he determined the centre of gravity of simple figures like triangles and discovered the relationship between weights and distances in relation to levers. 'Give me a point of support and I shall move the world,' he proclaimed, for he had shown that, provided the lever was long enough, any weight could be supported. With hydraulics he started with postulates such as 'Let it be granted that bodies which are forced upwards in a fluid are forced upwards along the perpendicular [to the surface] which passes through their centre of gravity.' From such postulates he proves that the loss of weight which a body experiences in water is equal to the weight of water displaced. Such principles explained why bodies float and enabled him to determine the specific gravity of gold and silver. No wonder he shouted '*Eureka!*' and leaped from the bath – it was a wonderful and totally surprising discovery. The application of mathematics to physical problems is in itself surprising, for it is far from obvious why the abstract language of mathematics should be able to provide so satisfying a description of the world.

Archimedes' work is a monumental achievement. Do most of us, lying in our baths, understand that our loss of weight is equal to the weight of water we displace? And that if this weight is greater than our own we will float? Could we tell whether a crown was made of gold or silver?

Cosmology is another area that demonstrates the triumph of Greek science. Every civilization and culture has provided its own answer to the question: what is the structure of the universe? Only Western civilizations, starting with the Greeks, have used studies of the heavens to provide an answer: other cultures have shaped their cosmologies on terrestrial events, the heavens merely providing an enclosure. For example, in one form of Egyptian cosmology the earth is depicted in some detail as an elongated platter – involving water, earth and air – an image probably taken from the Nile. The sun was the god Ra, who had two boats for his journeys across the skies – one during the day, one at night. Such cosmologies, it has been suggested, were not really meant as explanations but rather reflected the social structure of the society in which the people lived, and helped stabilize it. There is thus nothing in the Egyptian cosmology which even tries to account for Ra's journeys or their seasonal variations.

Ancient astronomers, such as the Babylonians and Egyptians, made many observations on the movements of the sun and stars, but these did not form part of an explanation. The Egyptians were primarily concerned with their use in establishing a calendar, while the Babylonians were interested in the accurate prediction of events in the heavens, such as the appearance of the new moon. The attempt to provide an explanation was first made by Anaximander, Thales's contemporary in Miletos, who assigned sizes to some heavenly bodies and likened the moon and its eclipses to the turnings of a wheel. With time, over two centuries, the Greeks developed a 'two-sphere' universe – the earth being a tiny sphere suspended at the geometric centre of a much larger sphere that carried the stars. This model has considerable conceptual elegance and provides, for the first time, an economical way of linking observations into a coherent whole. It is still convenient to use this model when learning navigation today.

Even in Greek times there were competitors to the two-sphere model. In the third century BC, for example, Aristar-chus proposed that the sun was at the centre and the earth revolved about it. But that clearly contradicted common sense, and Aristotle gave cogent arguments as to why the earth is the centre of the universe. For example, if the earth were moving in space we would surely sense it, and why would we not fall off? For Aristotle there was perfection in the heavens, and they contained the power on which terrestrial life depends. The authority of Aristotle's ideas derived in large part from his ability to express in an abstract and consistent manner a perception of the universe that embodied a spontaneous conception of the universe which had existed for centuries. They embody the ideas of many primitive tribes and children.

The concept of circular motion of the heavenly bodies about the earth created great problems when it came to under-standing the movement of the planets. Because they, like the earth, in fact rotate around the sun, their motion did not fit in with simple circular motion. In AD 150, Ptolemy provided the most comprehensive explanation of their complex motion in terms of epicycles, which had been proposed earlier – that a planet rotates about a small circle which in turn rotates about the earth. But the problem was that, while considerable accuracy in predicting planetary motion was achieved, this accuracy was at the price of complexity – more and more epicycles had to be added to fit planetary observations.

Given the apparent progress of Greek science, we are faced with a problem. Why did progress in astronomy and other sciences stop, effectively, until the arrival of Copernicus, Kepler and Galileo? Copernicus's ideas were, in principle, accessible to the Greeks, in the sense that they required no new observations. They were inaccessible, perhaps, because of the ideas themselves, which required a major conceptual advance. It was a barrier that is hard to explain. It was only with respect to the relative motion of the earth and planets with respect to the sun that Copernicus broke with tradition: all the rest of his work was in the Ptolemaic mould. Copern-

icus attacked Ptolemaic astronomy primarily because Ptolemy had not strictly adhered to the Aristotelian precept that
all heavenly motions must be explained by uniform circular
motions alone. While Copernicus's view of the heavens, with
the sun at the centre, was crucial for further advances, it was
not a simplification, for he made no real attempt to explain
the motion of the planets. That was left to Kepler, who more
than fifty years later made use of Tycho Brahe's observations
and set out to provide a physical cause for those movements.
Kepler had the intellectual courage to abandon motion in a
circle for motion along an elliptical path.

As with mechanics and motion, these new ideas rarely
relied on new observations but relied instead on a change in
thinking. In part the pervasive influence of Aristotle had to
be rejected, and this was much more difficult than it might
appear. Nothing illustrates this more clearly than Galileo's
analysis of falling bodies.

In Aristotle's view, the motion of an object up or down
was related to the object's natural place, and this in turn was
governed by its constituents – steam went up because it
contained fire; stones fell to earth because it was their natural
place, and the bigger the stone the faster it fell. Hence, according to Aristotle, the rate of fall of a body is proportional to
its weight. But, as Galileo says in the words of one of his
characters, Salviati, 'I greatly doubt that Aristotle ever tested
by experiment whether it be true that two stones, one weigh-
ing ten times as much as the other, if allowed to fall, at the
same instant, from a height of say, 100 cubits [the height of
Pisa's tower], would so differ in speed that when the heavier
reached the ground, the other would not have fallen more
than 10 cubits.'

Sagredo, Galileo's interested layman, reports that he has
done the experiment and it is not true, and Salviati continues,
'even without further experiment, it is possible to prove
clearly, by means of a short and conclusive argument, that a
heavier body does not move faster than a lighter one provided
the bodies are of the same material.' If, he goes on, we take
two bodies whose natural speeds of fall are different, it is
clear that, on uniting the two, the more rapid one will be
partly retarded by the slower, and the slower will be some-

what hastened by the swifter. But if this is true, and if a large stone moves with a speed of, say, eight, while a smaller moves with a speed of four, then the system will move with a speed of less than eight when the stones are joined. But the two stones when tied together make a stone larger than that which before moved with a speed of eight; hence the heavier body moves with less speed than the lighter – an effect which is contrary to the original supposition. Aristotle's ideas are wrong, since, if the rate of fall is proportional to weight, a logical contradiction can be demonstrated.

This delicious argument is an example of the kind of scientific thinking that was essential to the revival of science in the sixteenth and seventeenth centuries when Galileo was one of the giants. It also shows that one doesn't always need to do experiments to falsify a theory, though Galileo was a master of experimental method, and that internal consistency is one of the fundamental requirements of a scientific theory. The puzzle – unsolved – is why even though there were critics of Aristotle's ideas on motion, such as the Christian Stoic Philoponus in the sixth century, it took some 1,800 years for someone to point out the inconsistency or do the experiment. It requires a particular interest and mode of thought to deal with scientific problems.

It is particularly puzzling since Archimedes's method of reasoning compares very favourably with that of Galileo. It should come as no surprise that one of Galileo's first studies was in fact Archimedes. 'Those who read his works,' wrote Galileo, 'realize only too clearly how inferior are all other minds compared with Archimedes's, and what small hope is left over of their discovering things similar to the ones he discovered.' This, in my view, was not merely a fashionable enthusiasm for the past but a just assessment. What is remarkable is the maintenance over all those years of the Archimedean tradition. For this we must thank later Greek and Islamic scholars of the Middle Ages. A Flemish Dominican, Willem Moerbecke, in the thirteenth century translated every Archimedean treatise from the Greek into Latin, the language of scholarship at that time; and the Venetian editions of the sixteenth century were crucial in enabling Galileo to learn about Archimedes. Praise, too, to these noble scribes.

Why was it, then, that progress in science occurred in the West so many years after the Greeks? After all, each of the three inventions that Francis Bacon had identified as bringing about great changes in Renaissance Europe – printing, gunpowder and the magnetic compass – was a product of China, not Europe. The Chinese were brilliant engineers, but, though accurate observers of celestial phenomena, their contributions to science were minimal. They could build great bridges, and cast iron many years before it was done in Europe, but they never developed a mechanical view of the world. (Egypt provides us with another example of a highly sophisticated civilization which flourished for many centuries without making a single contribution to the development of the exact sciences.) The Chinese were fundamentally practical, but they had a mystical view of the world, a view which contained no concept of laws of nature but which was more directed to a social ethic whereby people could live together in happiness and harmony. Attitudes of this kind, in contrast to the passion for rationality that characterizes Christianity, perhaps partly account for science's flowering in the West and its failure in the East even to begin.

Even though the Chinese were the most persistent and accurate observers of celestial phenomena before the Renaissance, they did not develop a planetary theory and they did not have access to a geometrical theory. There was no Chinese Euclid. There were several classical astronomical cosmologies in China – that most commonly held took the planets and stars to be lights, of unknown substance, floating in an infinite empty space, and this fitted in with the infinities of time and space postulated by Buddhist thinkers, in which it took untold time for an object thrown from one Buddhist heaven to reach another.

The conception of the universe common to all Chinese philosophy is neither materialistic nor animistic: it is magical, even alchemical. The universe is viewed as being a hierarchically organized mechanism in which each part reproduces the whole. Man is a microcosm which corresponds with the whole universe – man's body reproduces the plan of the cosmos.

In religious Taoism, the interior of the body is inhabited by the same gods as the universe; there is a correspondence between the organs of the body and the holy mountains and seasons. The five organs of the body, its orifices and passions, correspond, for example, with the five holy mountains, the sections of the sky and the five elements – earth, fire, water, metal and wood. Understanding man thus provides, at the same time, an understanding of the universe.

Central to Taoism is the intimate link between nature and man, between human society and the universe, together with the idea of the cyclical nature of time. Thus all the ideas dealing with the nature of the world were intimately linked to man himself. In the Taoist view, all beings and things are fundamentally one. The opposing of opinions is disliked, because this involves a personal view and so loses sight of the whole. The person who wants to know Tao is told, 'Don't meditate, don't cogitate . . . Follow no school, follow no way and then you will attain to Tao.' Knowledge was to be discarded, for the ancient Taoist thinkers had an intense mistrust of the powers of reason and logic to develop anything resembling the idea of the 'laws of nature'.

Confucianism, the other dominant Chinese religio-philosophical tradition, is mainly about human conduct and much less about nature: it is about personal cultivation, aesthetics and purity. Confucian interest in cosmology was quite limited, though there were attempts to try to harmonize the traditional theories of Yin and Yang and the five elements. In one scheme they give birth to each other or overcome each other. It was essentially a numerological scheme linking man to nature – the four limbs being equivalent to the four seasons, for example, and the 366 bones corresponding to the days of the year.

These Chinese philosophical views contrast sharply with that of the Greeks, who had managed to distance man from nature. In fact it is almost universal among belief systems not influenced by the Greeks that man and nature are inextricably linked, and such philosophies provide a basis for human behaviour rather than explanations about the external world. These philosophies confine their curiosity to what affects man.

Could it be that these philosophical beliefs prevented the origin of science in their societies? On receiving a letter asking why it was that science arose only once, and in Greece, where it was confined to a tiny élite, and then persisted only in the West, Albert Einstein replied:

The development of Western science has been based on two great achievements, the invention of the formal logical system (in Euclidean geometry) by the Greek philosophers, and the discovery of the possibility of finding out causal relationships by systematic experiment (at the Renaissance). In my opinion one need not be astonished that the Chinese sages did not make these steps. The astonishing thing is that these discoveries were made at all.

It is true that the Chinese did not possess geometry, but perhaps their philosophy and the absence of capitalism were also important.

There was a long gap between the founding and flowering of Greek science and the next flowering in the time of the Renaissance. The interval between Archimedes and Galileo was almost 1,800 years. Archimedes and Euclid thrived at a time which coincided with the rise of Roman power. While the Romans were impressed by Greek culture, they were quite uninterested in the achievements of Greek science. Science and philosophy were relegated to a low status. Why science finally persisted in the West is not known but is the subject of extensive studies.

A case can be made for the importance of Christianity in fostering, in the West, the rationality, in the sense of logical arguments and reasoned discussion, that was necessary for science, and in also providing a system in which there was the possibility – even the conviction – that there were laws controlling nature. Such a conviction was unique to Christianity.

There is, however, a real methodological danger in looking for elective affinities between Christianity and science. If, for example, science had revived in some area of the world other than Catholic Europe, one would, I think, have little difficulty in explaining why Christianity *prevented* the rise of science – for, after all, Christianity is based on virgin birth and resurrection from the dead and relies heavily on untested dogmas – and one would find some other elective affinity to explain why it arose where it did. It is with such reservations in mind that the case for Christianity playing an important role in Western science must be considered.

The relationship between religion and science is an intimate one, but a most important aspect of the Christian religion alone is its role in supporting and fostering rational thinking. A key concept in Christian thought is that of order and law as manifested by God, the Creator. God, so it is claimed, even seems to approve of quantification: 'You have disposed everything in measure, number and weight' is quoted by St Augustine when discussing whether or not God knows all numbers including the infinite. Though St Augustine had serious doubts about the value of natural science, feeling no dismay if Christians were ignorant about the motion of the stars, he had no doubt about reason itself, which he not only held in high esteem but regarded as a divine gift.

Most important, early Christianity became enmeshed in metaphysical arguments which were closely linked to the nature of the physical world. Questions were asked about the purely physical nature of Jesus and how he could have two essential natures. Such questions gave rise to Arianism, a Christian heresy of the fourth century, which claimed that Christ is not truly divine but a created being; only God is self-existent and immutable. The relevance here is that logical consistency and reasoned argument, inherited from the Greeks, were important features of early Christianity.

The historian Pierre Duhem suggested that a decisive feature for the origin of science was the Christian refusal to accept the ancient pagan dogma of the divinity of the heavenly bodies. Failure to abandon such a view, according to which the sky determined events on earth, was perhaps even a brake on Greek science. But then, on the other hand, perhaps it was astrology – essentially non-science – that maintained an interest in astronomy.

Christian scholasticism was concerned with ideas like Being, Essence, Cause and End. It provided answers to the sort of questions that children ask – questions such as 'Who made the moon?' and 'Why . . . ?' The harmonizing of Christianity with Aristotle was due to Thomas Aquinas. In the middle of the thirteenth century, Arabian–Aristotelian science was disturbing the true believers in Europe: perhaps for the first time, Christian believers and theologians were confronted with the rigorous demands of scientific rationalism. An important influence was that of Averroës, the outstanding representative of Arabic philosophy in Spain. He regarded Aristotle's physics as divine and without flaw for the last 1,500 years. In order to avoid a contradiction between faith and reason while remaining true to Islam, he asserted that religious knowledge was entirely separate from rational knowledge as acquired in scientific studies. Opposing the spread of such ideas into Christianity, Aquinas held that theology is a 'science': it is knowledge that is rationally derived from propositions that are accepted as certain because they are revealed by God. Nature is quite distinct from man and has necessary laws. God exercises a supreme government over nature which conforms to the laws of a creative providence that wills each being to act according to its proper nature. Whereas nature cannot but conform to inalienable law, man has free will.

Aquinas treated motion as a branch of metaphysics and, following Aristotle, believed that everything that moves is moved by something else. God thus exists as the prime mover, for otherwise there would necessarily be an infinite regression of prior causal motions. By making Aristotle orthodox and compatible with Christian beliefs, a licence was given for cosmology to become a creative element in Christian thought, and permission for scientific thinking was granted.

The Christian picture of nature, especially as seen through Aquinas's eyes, is completely different from that of the Chinese. To take another example, fundamental to Buddhism is the idea of continual rebirth: that the whole world system goes through an inevitable process of growth, duration and destruction. Nothing, not even the gods, has a permanent existence. Reincarnation is a central feature. As the Catholic historian Stanley L. Jaki has said, 'That in all cultures – Chinese, Hindus, Mayan, Egyptian, Babylonian, to mention only the significant ones – sciences suffered a stillbirth can be traced to the mesmerizing impact which the notion of eternal returns exercised upon them.' Reincarnation goes against a set of laws governing nature in a causal way.

While emphasizing the contribution of Christian society to science, the contributions of Islam must also be recognized. Islamic scholars also continued the Greek tradition, and it may not be irrelevant that Islam offers a unifying perspective of knowledge and considers the pursuit of knowledge to be a virtue. It could, of course, not have been Christianity alone that was responsible for the flowering of science in the West in the sixteenth century.

Another way of considering the change in thinking that culminated in the scientific revival is related to economic factors. Max Weber has pointed out how 'every economic rationalization of a barter economy has a weakening effect on the traditions which support the authority of a sacred law'. And by 'rationalization' he means 'that there are no mysterious incalculable forces that come into play, but that one can, in principle, master all things by calculation'.

The idea of rationalization is at the core of Weber's concept of industrialization. While the concept is complex, a key aspect is the 'substitution of the unthinking acceptance of ancient custom, of deliberate adaptation to situation in terms of self-interest'. Perhaps it is this self-interest, then, that drives science, for self-interest is the best road to understanding, rather than vice versa. All this is intimately bound up with the development of capitalism. But, Weber argues, it was not just the capitalist spirit that drove science: the change in ethical outlook brought about by the Reformation was perhaps also essential, for the Protestant ethic encouraged a belief in progress and rational inquiry.

The view presented here is that only those societies influenced by the Greeks developed science. Is this really so? Some anthropologists have argued that the beliefs of so-called primitive peoples are similar to those of science. In African cosmologies, for example, the gods of a given culture are part of a scheme which helps interpret the diversity of everyday life in terms of the action of relatively few kinds of 'forces'. These 'forces' are not really causative agents but are the result of the activities of life ancestors, heroes, water-spirits and so on. And indeed diviners and witch-doctors do provide explanations of a causal kind which are what one might call 'common-sense', or mythological, explanations. All the explanations are related to human conduct, not to curiosity about nature itself, and they are, of course, devoid of all the key characteristics of science.

In these traditional cultures, the possibility of alternative concepts or mythologies is simply absent: the system is closed. The Azande tribesman, for example, cannot think his thought is wrong. Because his web of belief in relation to witch-doctors and oracles is so tightly woven, it is the only world he knows. Failures of prediction by witch-doctors are 'excused' by making use of other beliefs. Nothing that occurs, even repeated failure, will be used as evidence against the belief system. The Azande have the convictions of any religious group whose beliefs cannot be altered by secular experience (see Chapter 6).

An interesting feature of traditional African communities is the complete lack of admission of ignorance about some question which the people themselves consider important. So, while they might admit to not knowing where the world comes from, this is primarily because it is a question without interest. For questions relating to disease or crops, this is never the case, and an explanation is always offered. Indeed, no important event ever passes without an explanation. If someone is killed by accident – by a falling branch of a tree, say – there has to be a definite explanation, framed in terms of revenge or sorcery. The idea of chance and unpredictability is not acceptable. This, of course, precludes any possibility of thinking in terms of probabilities, which play an important role in scientific thinking.

The lack of curiosity about natural phenomena is very relevant, for it was self-conscious questioning that made Thales and the Greeks both unique and so important. Indeed, one of the characteristic features of magical thought that makes it so different from science was made clear by Keith Thomas in *Religion and the Decline of Magic:* once initial premises about the nature of the world are accepted, no subsequent discovery will break the believer's faith, for he can explain it away in terms of the existing system. This is an important statement of how science does *not* proceed. A Greek poem by Agathias (AD 536–582) about a farmer,

Kalligenes, who consults an astrologer about his crops, illustrates this nicely:

> The astrologer cast his stones across the board,
> Studied them, wiggled his fingers and said:
> 'If, Kalligenes, there is rain enough
> On enough of your land, and if the weeds
> Don't take over, nor frost wreck the lot,
> If a hailstorm doesn't knock it all flat
> If the deer don't nibble, if no calamity
> Up from the earth or down from the sky
> Occurs, the signs show a good harvest
> Unless there's a plague of grasshoppers.'

So how did thinking break out of this mould? Perhaps the answer lies in the ideas about religion and rationality provided above. But what is interesting is that it is unlikely that it was science itself that caused the decline in magic.

According to Thomas, magic was on the decline before the rise of science and technology in the Renaissance. How otherwise, for example, can one account for the fourteenth-century Lollards, a religious sect who renounced the Church's supernatural protection against disease or infertility yet had nothing with which to replace it? Even in the seventeenth century the decline of magic in relation to medicine was not due to an improvement in treatment – William Harvey's discoveries about the circulation of the blood (which, ironically, were made in the context of Aristotelian thought) did nothing practical for medicine. It can even be argued that medical innovations based on science did little to prolong life until the nineteenth century.

Perhaps for most people, then and now, it is authority in the form of education and received opinions that determines their basic assumptions about science, religion and magic. As the psychoanalyst Ernest Jones so aptly wrote, 'The average man of today does not hesitate to reject the same evidence of witchcraft that was so convincing three centuries ago, though he usually knows no more about the true explanation than the latter did.'

Unless one has a specific self-aware and self-critical curiosity, even basic biological principles can be ignored. The anthropologist Ashley Montagu claimed that among some Australian aborigines, although intercourse is in some way associated with pregnancy, it is generally considered only to be one of the conditions, not the cause. 'The effective cause of pregnancy, *and nothing else*, is the immigration into a woman of a spirit-child from a specifically known external source, such as a totem centre, an article of food, a whirlwind and the like.' Such beliefs do not represent conclusions arrived at in the course of discussion or reflection. They are, however, beliefs which are repeatedly verified: they work, therefore they are true. Girls marry at puberty and will not bear children before marriage. Intercourse is frequent, yet some girls have babies and others do not. Thus the relationship between intercourse and childbirth is far from obvious. However other anthropologists have argued against the idea that some primitive people are ignorant about the relationship between intercourse and childbirth, since these people have experience of animals. On the other hand, it is worth pointing out that even at the beginning of the nineteenth century in Britain the biological function of intercourse was not understood. There was a deep conviction that the sperm alone was responsible for creation of the embryo – a belief dating back to Aristotle. For this reason, 'Every naturalist, and indeed every man who pretended to the smallest portion of medical science, was convinced that his children were no more related, in point of actual generation, to his own wife, than they were to his neighbour.' The mammalian egg was only discovered in 1827.

Persuasive evidence for the unnatural nature of science is that for thousands of years the mythology and cosmology of almost all cultures entertained neither a critical tradition nor curiosity about nature. The idea that man is innately curious is a partial myth: man's curiosity extends only to what affects his conduct. How otherwise can one explain the widespread lack of curiosity about nature for its own sake in society after society? And the historical perspectives I have offered support a view of the evolution of science as a very chancy affair. Like the evolution of life itself, there needed to be very special conditions for science to have started at all as it did in Greece. (The same is true for the alphabet, which also had a unique origin.) In *Wonderful Life*, Stephen Gould has emphasized the role of contingency – accident – in biological evolution. Considering human evolution, he writes, 'Arguments of this form lead me to the conclusion that biology's most profound insight into human nature, status and potential, lies in the simple phrase, the embodiment of contingency: *Homo sapiens* is an entity, not a tendency.' If, he argues, the tape of evolutionary history were to be rerun with slight differences which let *Homo sapiens* expire in Africa, then *Homo sapiens* might never appear again. There is no guarantee of progress towards human beings of our form or potential. If the dinosaurs had not mysteriously disappeared – a cosmic catastrophe perhaps – and been replaced by mammals, we would not be here. 'In an entirely literal sense, we owe our existence as large and reasoning mammals to our lucky stars.' And so it is with science. There was no inevitability that science should have arisen in Greece, or that it should have been sustained by the intellectual climate of Christianity and capitalism. It is not clear whom we should thank – perhaps the planets again, because thinking about them undoubtedly played a major role in the history of science. Without the planets there would, for example, have been no Copernicus. But it is always to Miletos and to Thales that it is necessary to return. We must thank him and his contemporaries.

4
Creativity

Among the confusions about the nature of science is a widespread attachment to the idea that arts and sciences are basically similar in that they are both creative products of the human imagination, and that attempts to draw a dividing-line are quite wrong. Even scientists tend to share this view, and the great German physicist Max Planck asserted that the pioneer scientist 'must have a vivid intuitive imagination, for new ideas are not generated by deduction, but by an artistically creative imagination'. A similar line was taken by Jacob Bronowski in *The Common Sense of Science:* 'The discoveries of science, the works of art, are explorations – more, are explosions, of a hidden likeness. The discoverer or the artist presents in them two aspects of nature and fuses them into one. This is the act of creation, in which an original thought is born, and it is the same act in original science and original art.' This view, however, is misleading and possibly sentimental. Scientists are, of course, creative, and do require a 'vivid intuitive imagination', but their creativity is not necessarily related to artistic creations. It is only at a relatively low level that creativity in the arts and in science may be similar: a level which includes almost all human activities that involve problem-solving, from accountancy to tennis.

Differences between creativity in the arts and science reflect some of the important differences between the two. Creativity in the arts is characteristically intensely personal and reflects both the feelings and the ideas of the artist. By contrast, scientific creativity is always constrained by self-consistency, by trying to understand nature and by what is already known. How unlike the French novelist Alain Robbe-Grillet's view of the novel which 'crosses itself, repeats itself, bisects itself, contradicts itself'. Moreover, the scientists' 'creations' ultimately become assimilated into public knowledge, as in textbooks, and their contributions are, with rare exceptions, ultimately anonymous (Chapter 5). With artists it is the original creation that is all-important. Even more significant is the nature of what is created. A work of art is capable of many readings, of multiple interpretations, whereas scientific discoveries have a strictly defined meaning. Again, artistic creations may have strong moral overtones, whereas science, in principle, is value-free (but see Chapter 8). In addition to being personal, artistic creations are about singular, often internal, experiences, whereas scientists strive for generality and are interested, for example, in ideas that apply to all cells rather than just to particular ones. Whatever the scientists' feelings, or style, while working, these are purged from the final work. Finally, there are objective and shared criteria for judging scientific work, whereas there are numerous interpretations for artistic creations and no sure way of judging them. Given all these differences, one should treat claims for similarity between scientific and artistic creativity with deep suspicion.

Consider the mathematician Henri Poincaré's attitude to beauty:

The scientist does not study nature because it is useful; he studies it because he delights in it and because it is beautiful. Of course, I do not speak here of that beauty which strikes the senses, the beauty of qualities and appearance; not that I undervalue such beauty, far from it, but it has nothing to do with science. I mean that profound beauty which comes from the harmonious order of the parts and which a pure intelligence can grasp.

Scientific beauty is not easy to define, but it is related to simplicity, elegance and above all the surprise in finding a novel way of doing an experiment or a theory which explains things in a new way.

There are many styles in science, and scientific creativity comes in many forms; it is not found only in new ideas like those of Newton or Darwin. In some cases great advances have been made by designing a new apparatus for experiments, like the cloud chamber for observing the collisions of subatomic particles; in others, the brilliance lies in designing the experiments and carrying them out. In all cases the advances are underpinned by an imaginative conceptual framework. But it is no use for anyone to pretend that there is, at present, any real understanding of the creative process in any human activity. For example, the ideas about creativity offered by psychoanalysts are not about the creative process itself but are rather about the supposed reasons why men like Kafka, Newton and Einstein pursued the intellectual life. There is, for example, Anthony Storr's claim, in *The Dynamics of Creation*, that creative activity is an 'apt way for a schizoid individual to express himself'. Whether or not it is true, our understanding of the origins of Newton's or Einstein's genius is helped not one whit by their being men who related poorly to others. Paul Valéry's claim about Racine is equally true of Newton or Darwin: 'Collect all the facts that can be collected about the life of Racine and you will never learn from them the art of his verse.' At most one hopes for a glimmer of how their minds worked.

Even though our understanding of creativity is severely limited, it is possible to explore some of the ideas proposed to account for the origin of scientific ideas and to examine some case histories, since these also help to illuminate the process of science.

A widely held view is that creativity in science is based on what is known as evolutionary epistemology or the chance-permutation model. In essence, this model suggests that scientists randomly generate theories, of which the good ones survive since they are selected because of their explanatory powers. The creative process is said to entail mental elements which are permutated in a random manner, and these random permutations are selected by another process so that the best ideas survive. This is an approach which has a long history, since Descartes regarded it as a matter of indifference how scientific hypotheses were produced: the important point for him was to make hypotheses and to see where they led. He drew an analogy with deciphering a coded message, where by experimenting with certain substitutions one can eventually obtain the correct cipher even if the substitutions are chosen at random. A hypothesis was, in his view, to be judged by the usefulness of the conclusions that could be drawn from it.

Dorothy Sayers has, in modern times, expressed this idea clearly. Listen to Lord Peter Wimsey:

I always make it a rule to investigate anything I feel like investigating . . . but this is the real sleuth – my friend Detective-Inspector Parker of Scotland Yard. He's the one who really does the work. I make imbecile suggestions and he does the work of elaborately disproving them. Then, by a process of elimination, we find the right explanation, and, the world says 'My God, what intuition that young man has!'

Successful as Wimsey may be, this approach, in its extreme form, can be thought of as being no more than making use of Darwin's apes. If apes sat at word processors, randomly tapping the keys, then, in the course of time and provided we could recognize the good and important ideas, out would pop the theory of evolution, Newton's mechanics, the theory of relativity and all other scientific theories. This gives no insight into what is involved in generating ideas, for the question is: where do the random thoughts come from, are they really random, and is there no real creativity in the generation of the variations itself? It is silly to think that any one thought is equivalent to any other, and that every idea has an equal chance of being put forward. The mark of a good scientist lies precisely in the new variants proposed. In all branches of science there is a great deal that must first be learned and understood at a deep level, so that the right questions are posed, before the generation of new thoughts can even be contemplated. The number of scientists in a particular field at that level of competence is probably small: there is a strong selective pressure before anyone even enters the creative arena. The talent, gift, genius of scientists is first to understand properly the current state of a field, then to recognize what problems can be solved, then to generate creatively new ideas. The thoughts are not random, but that is not to say that they don't explore a wide range of new ideas, including some that at first sight may seem to be absurd. What is so impressive about good scientists is the imaginative solutions they come up with. Perhaps the analogy is with chess – choosing the right line many moves ahead: to think of the chess master as making random searches, like a crude computer program, is quite misleading.

Once we get rid of the random element in generating new ideas, however, we may be left with an important idea: the idea of bold conjectures, or guesses, followed by verification or falsification. For example, the molecular biologist Sydney Brenner has commented:

For twenty years I shared an office with Francis Crick and we had a rule that you could say anything that came into your head. Now most of those conversations were just complete nonsense. But every now and then a half-formed idea could be taken up by the other one and really refined. I think a lot of the good things we produced came from those completely mad sessions. But at one stage or another we have convinced each other of theories which have never seen the light of day . . . I mean completely crazy things.

The physicist Richard Feynman considered science to proceed by guesses:

In general we look for a new law by the following process. First we guess it. Then we compute the consequences of the guess to see what would be implied if this law that we guessed is right . . . If it disagrees with experiment it is wrong. In that simple statement is the key to science . . . It does not make any difference how smart you are, who made the guess, or what his name is – if it disagrees with experiment it is wrong . . . It is true that one has to check a little to make sure that it is wrong . . .

It is no shame to be wrong, only disappointing. But what Feynman does not point out is that some guesses are very much better than others, and he ignores the influence of doing experimental work. Even so, his approach is similar to that adopted by Newton.

Analysis of Newton's procedures has shown that they fit quite nicely with Feynman's guessing model. Newton's procedure in his *Philosophiae Naturalis Principia Mathematica* involves an alternation of two phases or stages of investigation. In the first, the consequences of an imaginative construct are determined by applying mathematical techniques to the initial conditions. In the second phase, the physical counterpart of the initial conditions or the consequences are compared or contrasted with observations of nature. The first stage removed constraints – Newton could explore the consequences of any consequences he found mathematically interesting. He explored the implications for planetary motion of Hooke's suggestion that bodies attract each other without concerning himself about the nature of the attracting force. Only later, when he had his celestial mechanics worked out, did he then turn his attention of the problem of the force.

There can be little doubt that bold, almost unconstrained, thinking can be an invaluable procedure. But, as always, the question of where the imaginative ideas come from is left unanswered. Nevertheless, this discussion should at least have dispelled the notion that progress in science comes only from the patient accumulation of facts and tedious observation. It is to the philosopher Karl Popper's great credit that he has emphasized the imaginative nature of scientific thinking.

In contrast to conscious guessing, unexpected, unconscious illumination is also regarded as a typical feature of scientific thought. The classic incident is that related by Poincaré, in relation to his solving a mathematical problem:

Then I turned my attention to the study of some arithmetical questions without much success and without a suspicion of any connection with my preceding research. Disgusted with my failure, I went to spend a few days at the seaside and thought of something else. One morning walking on the bluff, the idea came to me, with just the same characteristics of brevity, suddenness and immediate certainty, that the arithmetic transformations of indeterminate ternary . . . quadratic forms were identical with those of non-Euclidian geometry.

And an important advance in mathematics had been made.

A similar experience is related by the English mathematician Christopher Zeeman. Seven years after trying to prove a theorem in topology, that one could tie a sphere in a knot in five dimensions,

I sat down one Saturday morning and I thought 'Well I'll have another crack at this damn problem.' And lo and behold, I suddenly found to my surprise, that I had proved the opposite . . . and I was so excited that I spent the whole weekend writing this paper up, about twenty pages. And then late that night, I confess, I went and sat on the lavatory and while I was there the real flash of inspiration struck me like a bomb. I suddenly saw how to reduce the proof from twenty pages to ten lines.

It is not only in mathematics that such insights apparently come suddenly. 'A film of no importance. Slumped in my seat, I dimly perceive in myself associations that continue to form . . . I am invaded by a sudden excitement mingled with vague pleasure. It isolates me from the theatre from my neighbours whose eyes are riveted on the screen. And suddenly a flash. The astonishment of the obvious. How could I not have thought of it sooner'. Thus François Jacob's Nobel-Prize-winning insight into the essential similarity between how enzyme synthesis is turned on in bacteria and the replication of bacterial viruses; namely, that both are controlled by a special molecule binding to the DNA.

Attractive though unconscious processing of ideas might be, for it has a certain romantic ring of artistic genius à la Coleridge, the evidence that any real processing, testing of combinations of ideas, occurs in the unconscious has been questioned. What real evidence is there for novel thoughts coming via the unconscious? In every case where scientific illumination occurs suddenly, it is preceded by a long period of intensive conscious study. The need for rest, for a new start, may give a false impression of sudden discovery, for we can carry only a small number of concepts in our minds; when the problem-solver takes a rest from the problem for a time, information in the short-term memory that is not found to be contributing to a successful solution may be lost – selective forgetting. When the problem is returned to, a quite new path may be followed. There is even reason to doubt Coleridge's story of how he created Kubla Khan, which was, he claims, written in an opium trance and was interrupted by a person from Porlock who had come to discuss business. Again, the classic and influential story of Kekulé's dreaming about snakes biting their tales leading to the discovery of the six-carbon-atom benzene ring, of great importance in chemistry, may be less dependent on dreaming than he would have us believe. His injunction 'Let us learn to dream, gentlemen' may be unwise advice, for such insights are far from typical and are invariably dependent on an enormous amount of earlier work and preparation. For example, when Crick and Watson solved the DNA structure, the solution did come quickly at the end, but it was the result of a long process of hard work. And many other discoveries are far less dramatic.

Poincaré was, for a scientist, unusual in that he gave a great deal of thought to the nature of creativity. His own work pattern comprised a number of stages: conscious work, unconscious work, illumination (when he was successful) and then the work of 'verification'. Poincaré himself admitted that what he called unconscious work was always preceded by periods of conscious work. Poincaré also held to something like the random generation and selection view of creativity. But, as he rightly asked, how does selection occur, particularly if it is an unconscious process? His answer is not very helpful, since he talks of his 'intuition' guiding the choices in extremely subtle and delicate ways, that they are felt rather than formulated, and that the process also involved a sense of beauty.

Other scientists too gave much credit to the unconscious, and the nineteenth-century German physicist Von Helmholtz quoted Goethe's words:

> What man does not know
> Or has not thought of
> Wanders in the night
> Through the labyrinth of the mind.

It is hard to avoid thinking that intuition (and the unconscious) as used by Poincaré and others is no more than a convenient black box which contains the creative process but about whose workings we are ignorant. (Unfortunately, cognitive psychology, with its emphasis on connections, networks and computer programs, is no more illuminating.) But it is important not to confuse intuition as defined in this context with that used in our day-to-day lives. For example, as Einstein pointed out, a scientist's intuition rests on a technical understanding of what can be regarded as reliable and important. Common-sense intuition is quite different. While both are based on experience, the nature of that experience is very different. Scientific intuition relates not to common-sense experience but to the great fund of highly specific knowledge that has been acquired; it involves knowledge of how other scientists have solved problems, of what is expected of a scientific theory and of what may and may not be solvable. So strong was Einstein's conviction that he didn't have the necessary intuition that he decided not to become a mathematician: he knew he did not have the 'nose' to decide which were the really important problems.

In stark contrast to the claim for the scientists' imaginative-artist approach to creativity is that of the Nobel laureate in economics Herbert Simon and his colleagues. They believe that scientific creativity can be carried out by a computer program: that there is thus no real difference between the work of the 'genius' scientist and that of those of lesser ability, and so the idea of high creativity is a myth. For them, the process of discovery can be described and modelled.

Their central hypothesis is that the mechanisms of scientific inquiry are not peculiar to that activity but can be analysed as special cases of the general mechanisms of problem-solving. They do recognize science as a social process and also, since its goals when beginning to tackle a problem are usually not clearly defined, that it differs from ordinary problem-solving: finding problems and formulating them in a precise form is an integral part of science. In contrast, problem-solving, it is suggested, can be considered within the framework of cognitive psychology in terms of creating a symbolic representation of the problem and using operators on this. The search for a solution is not random trial and error but is guided by rule of thumb – by heuristics. For example, there are 50 billion billion (50×10^{18}) possible settings of ten dials on a safe if each is numbered from 0 to 99, but a click at the correct setting for each reduces the number of trials to open the safe to about 500. Good scientists merely have better heuristics and do not require 'intuition'.

Simon and his colleagues' major claim is to have developed computer programs which, using their problem-solving approach, can make discoveries over a wide range of topics. 'Discoveries' is not really the right word, for they have not discovered anything new; rather they have shown how the computer could have discovered universal gravitation from the information available to Newton, or Planck's constant – a fundamental quantity in quantum mechanics – given the information available to Planck, so all their demonstrations have involved the invaluable wisdom of hindsight.

In their programs, the criterion for the proposed solution is that the law found should fit the data 'well enough' – not worse than 3 per cent error. They claim that the generalization that the computer has found to fit the data will never be unique. Their approach is to ignore the small error and catch the 'rabbit' first. For them, the function of verification procedures is not to provide scientists with unattainable certainty or uniqueness for their discoveries but to inform them about the risks they are running in committing themselves to hypotheses that have been formulated and to provide guidance that may enhance their chances of making relatively durable discoveries.

While their computer programs may be successful, there have been criticisms, not the least being the amount built into the programs since the programmers do know the answer themselves. Their programs have made no new discoveries. But probably a more serious criticism is that scientific research involves more than just problem-solving: there is also data-gathering, description, explanation and theory-testing. The invention of new instruments, for example, does not fall within their computer programs: they are concerned only with 'the induction of descriptive and explanatory theories from data'. While they do recognize the importance of correctly formulating a problem, they claim that this is nothing more than a variety of problem-solving, a claim which is strongly disputed. One only need recall that Einstein's discovery of the theory of relativity was influenced by his posing the following problem: what would be the consequences of running alongside and then catching up with a point on a light wave? Computers couldn't 'think' like this.

For the choice of problem is crucial. As the Nobel laureate Peter Medawar put it, science is the 'art of the soluble', and part of that art is choosing a problem which will turn out to be soluble. Francis Crick, for example, makes much of this point in relation to protein structure. The early heady days of molecular biology led to the sequence hypothesis, namely that the sequence of amino acids in proteins, which was specified by the DNA, completely determined the properties of the protein (Chapter 1). It was known that, although proteins were synthesized as a linear chain of amino acids, this chain then folded up spontaneously into complex shapes. The three-dimensional shape adopted by the chain was fundamental to the protein functioning properly, for the special properties of proteins are due to the shape of the folded chain which gives each protein a unique configuration and determines its function. Now, although the sequence of amino acids was assumed to be sufficient to determine the folded structure, this had not been formally established, nor was it possible to predict the structure from the sequence. Crick and his colleagues decided not to tackle the protein-folding problem, although it was in many ways the obvious next step in their research. How right they were, since thirty years later this still has not been fully solved – it is an exceptionally difficult problem.

Another aspect of problem-solving which is beyond current computer programs is knowing when to approximate, which comes only from experience. Approximation involves making simplifying assumptions which will make a problem tractable – at the risk, of course, of oversimplifying and thus making the solution of less value. And perhaps of no less importance is to know when to stop working on a problem or to abandon a particular line of investigation. It can be painful to give up much past investment of effort and take a new line.

Whether or not Simon's problem-solving approach is correct – and I doubt that it is – it nevertheless contains an important idea, namely that at least part of scientific thinking is a kind of problem-solving of a very structured kind which can be simulated by computers. This emphasizes again the unnatural nature of scientific thinking, for computer programs of the type Simon and his colleagues use are quite unlike common-sense thinking. In general, computers are very bad at simulating common sense and such human activities as recognizing handwriting, and when they are successful in such activities they use a quite different mechanism to that used by our brains. But whatever talents computers may have, genius is not one of them.

Genius is always fascinating, raising some scientists to demigod status in the eyes of other scientists, but its nature remains an almost total mystery. Genius is usually judged with hindsight, but the scientific genius exerts a massive effect on both contemporaries and posterity. But no matter whether in any particular case the accolade of genius is applied or deserved, anyone who works in science quickly recognizes the leaders and truly creative workers in the field – they are often faster, more hardworking, more imaginative, cleverer, know more, understand more, speak better, calculate faster or possess at least some of these attributes.

Scientific genius can be recognized by the leadership the scientist gives and, more important, by the enormous influence on both contemporaries and posterity. Newton, Darwin and Einstein clearly qualify. Genius is ultimately ascribed for enduring eminence or reputation, which ought to reflect a contribution which has illuminated the nature of world. There may be some justification (Chapter 6) for regarding 'genius' as a social construct – social forces acting to establish who should be rewarded – but, while genius may include a social component, it is also about objective achievement. And while we may dispute whether someone is or is not a genius, we usually have little difficulty in identifying the outstanding scientists in any field.

Scientific genius also is quite different from that in the arts. One fundamental aspect that makes it different from genius in any other field is that, because science is a communal effort, in the long run the existence of scientific geniuses may be irrelevant: given time, resources and a sufficiently large trained community committed to science, all discoveries will probably be made. (In fact, the Ortega hypothesis, discussed later, claims that experimental science has progressed largely through the work of those of mediocre talent.) Discoveries and progress in science need not depend on any single person: if not Newton, then many others; if not Einstein, then X, Y and Z. The pathway of progress may be different, but in the end the result will most likely be the same. Not so for *Hamlet* or *Così fan tutte* – there are no replacements for Shakespeare and Mozart. Again, the description by François Jacob of scientific genius emphasizes further differences:

In science the great man is, first of all, the one who knows how to spot the right problems at the right moment, while there is a chance of solving them. He is the one who knows how to surround himself with the right collaboration, to find among his pupils those capable of becoming his successors, and of developing the theories he has set forth . . .

Hardly an apt recipe for a Shakespeare, a Mozart or a Picasso.

The psychologist Howard Gruber has considered the problem of how early gifts and talent are transformed into exceptional creativity. It seemed to him that Thomas Henry Huxley was more brilliant and versatile than Darwin, his contemporary, and that any committee looking at their early research plans – Huxley's for the voyage of the *Rattlesnake* and Darwin's for the voyage of the *Beagle* – would have favoured Huxley. And when Huxley heard of Darwin's theory he exclaimed, 'Why didn't I think of that?' Why indeed? And the same is probably true for many others too. Gruber's answer to Huxley's question is: openness – young Darwin's vague and open receptiveness was more successful than Huxley's hard-edged analytic approach. But this is not necessarily a universal formula for success. How are we to understand the difference between the gifted and the extraordinary?

Studies on intellectually and academically gifted children show them to be highly efficient in the use of both short- and long-term memory processes. However, a more important feature is that they are capable of what are called metacognition and metamemory. Metacognition refers to a person's awareness of his or her thought processes and is assumed to be crucial for the selection and implementation of complex problem-solving strategies. Metamemory is a subcomponent of metacognition and again refers to self-knowledge, in this case about the person's memory system. Gifted children were found to be particularly aware of the strategies they used for remembering, such as interest in the topic and how they linked up thoughts. Successful scientists are similarly self-aware – it is one of their defining characteristics.

Studies on skill at chess may help to illuminate some aspects of skill at science. When grandmasters were compared with experts, there appeared to be no difference in the approach to chess problems and in their skill in solving them. The difference is due to motivation, character and knowledge. The grandmaster has a richer knowledge-base to draw on, due to thousands of hours of play and study. One needs passion and discipline to devote such time and energy, and that is a matter of character. There is a strong sense of truth in this when applied to gifted scientists: they have stamina, devotion, psychic courage and 'character', and they work very hard at problems.

The cry of *'Eureka!'* may be rarer than popularly supposed, but, even so, the cry does ring out over the centuries. But the cry is often misleading, for it suggests that the solution to a scientific problem comes in a moment of divine, or ablutional, inspiration; it neglects the slow and often painful process from the formulation of the problem, through false turns, to that lovely moment of solution.

These points are encapsulated by Newton's reply to the question of how he had discovered the law of gravity: 'By thinking on it continually.' Gruber makes the point that science, and creativity in general, has a long time-scale – the ideas of Newton, Darwin and Einstein took many years to develop. If there is a blinding flash, a *'Eureka!'*, we should not forget the 'years' that had previously been spent in thinking about the problem. Are they really less important? It is a characteristic feature – almost a defining feature – of science that it takes a long time to solve a problem. This is partly due to the difficulty inherent in the problem and partly because science is social and it is necessary to learn what other scientists have done in order to assimilate current knowledge.

Among the important characteristics of the great scientist is the ability, already referred to in relation to Crick, to recognize which problems to solve. It is also important to recognize which evidence and ideas to accept or discard. To challenge well-established beliefs can be remarkably difficult intellectually. One has also to make hard judgements about the available experimental data: as Francis Crick has pointed out, a theory that fits all the facts is bound to be wrong, as some of the facts are themselves bound to be in error (see Chapter 5).

The detailed study by Gruber of the origin of Darwin's ideas on evolution provides a valuable case-study of scientific creativity. One of the first important ideas that Darwin developed was in relation not to evolution of animals but to a geological problem, the formation of coral reefs. While on the west coast of South America in 1835, Darwin put forward the idea that coral reefs were formed by the upward growth of the coral during the sinking of the land. This was, in a way, an evolutionary theory, in that it required the limitation of growth of the corals – corals do not grow beyond a limiting distance above the water – and the theory explains why the variation in forms of coral islands is continuous.

We share, however, Gruber's surprise that Darwin's first theory of animal evolution was along somewhat strange lines. In order to account for species changing and yet being adapted to their environment, and yet also for the number of species remaining approximately constant, Darwin invoked the emergence of simple life-forms, called monads, by spontaneous generation. The monads, he suggested, evolved as the result of direct environmental influence but had only a limited lifespan, so the species to which the monads gave rise eventually died and became extinct.

It is only possible to understand this seemingly ludicrous idea in terms of the concept of species present at that time. Each species was thought to contain its own specific essence, and thus it was impossible to imagine that it could either change or evolve. The ideas of the geologist Charles Lyell, who greatly influenced Darwin, illustrate this very clearly, since Lyell could not conceive that one species could be converted into another. And, if Lyell, who was so close to evolutionary thinking, could not conceive of this, then it was even less acceptable for his predecessors such as Lamarck. Lyell's criticism of Lamarck was severe, particularly in respect of Lamarck's ideas that there was a progression towards perfection. Lyell realized that species were the key and pondered about how they arose and became extinct without making real progress. Unlike Lamarck, he believed that species could become extinct either through physical factors or – and this was significant – through competition with other species. But as to the origin of new species he could merely say, 'Species may have been created in succession at such times and at such places as to enable them to multiply and endure for an appointed period . . .' This was a doctrine of special creation, and Darwin's monads are in a similar line of thought, though later, in the *Origin of Species*, Darwin gave much attention to rejecting special creation.

It is all very well to write computer programs that will arrive at the idea of evolution by natural selection – hindsight gives wonderful wisdom – but such programs take no account of the paradigms or ideas of the time, such as the idea that species are immutable. It is often difficult to recognize how hard it was to break with current concepts.

The very title of the notebook in which Darwin wrote down his first evolutionary theory in July 1837 is in itself revealing – *Transmutation of Species*. From the beginning, he writes, 'Each species changes . . . The simple cannot help becoming more complicated.'

In his new monad theory there was a crucial innovation: the idea of branching evolution, the tree of life. 'Organized beings represent a tree, irregularly branched.' But the branching model of the monad theory required the simultaneous extinction of many species, which is implausible. Darwin thus began to consider the possibility that monads have a variable lifespan; but he recognized the weakness of the idea, and by September 1837 the monads had died. He had in one way abandoned the problem of the origin of life, and this had the enormous virtue of simplifying what was already a very difficult problem. In the summer of 1838 this is made explicit, since he enjoins himself to not to try and go too far back 'for if so it will be necessary to show how the first eye is formed – how one nerve becomes sensitive to light . . . which is impossible.'

In the monad theory, with its branching tree, variation arose from an inherent tendency to progress; but this, Darwin realized, provided no explanation of its mechanism. In fact, the origin of variation was forever a problem for Darwin, for genetics was still to be developed. Why there should be variation in animal size and shape was simply unknown. He was very conscious of this gap in his theory, and he showed great intellectual courage – Gruber calls it heroic – in basing his theory on a mechanism he was unable to explain. He did, however, put forward a theory along Lamarckian lines in which acquired characters could be in principle inherited.

Here there is an important similarity with Newton, who stuck to gravity without having an explanation of its underlying mechanism. Both Newton and Darwin were driven by the data and were forced to recognize that they couldn't explain everything. It may be a characteristic of great scientists to know what to accept and what to leave out.

In a famous passage, dated sometime in 1836, Darwin at last doubts the stability of species: 'When I see the Islands in sight of each other, and possessed of but a scanty stock of animals, tenanted by these birds, but slightly differing in structure and filling the same place in Nature, I must suspect they are only varieties . . . such facts would undermine the stability of Species.' A similar idea opens the first *Transmutation of Species* notebook: 'According to this view animals on separate islands ought to become different when kept long enough apart . . .' It was not until March 1837 that Darwin fully appreciated the significance of the island fauna, when the ornithologist John Gould told him of the distinctness of the hummingbirds in his collection which came from three different Galapagos islands.

Darwin was also impressed by recent work that had shown that micro-organisms could reproduce extremely rapidly. Then a few days later he read Malthus's *Essay on Population*, with its emphasis on the enormous over-productivity of nature without any checks on fecundity, and 'it at once struck me that under these circumstances favourable variations would tend to be preserved, and unfavourable ones to be destroyed. The result would be the formation of new species.' Thus was born the theory of evolution by natural selection. Reading Malthus enabled him to realize that natural selection would not only select out non-adapted variants but would also favour those variants that were better adapted than others.

It is quite clear that Darwin's theory required a long incubation period and many false starts. It also required fine judgement, great persistence, intellectual courage and, finally, genius. The path towards the solution had not been straightforward.

Gruber emphasizes that 'attacking the most difficult tasks requires the highest level of aspiration, and consequently puts stressful demands on the ego system.' There must be a sense of special mission, and also a degree of psychic courage in taking on a very difficult problem in such a way that if the project fails there is nothing to show for it. There are probably many examples of this, but most scientists adopt a safer strategy, such that *something* positive will come out of the research. For example, if Watson and Crick's attempt to determine the structure of DNA had failed, they would have had very little to show for their efforts.

Another example is the American molecular biologist Mark Ptashne's search for the repressor protein. In 1965, at the age of twenty-five, Ptashne decided to try to isolate the key repressor protein that had been postulated by the model of the French biologists François Jacob and Jacques Monod. This protein had been postulated to bind to a specific region of a gene in a bacterial virus and in so doing to play the crucial role of switching a gene off. Control of gene activity, that is turning genes on and off, is fundamental to cell behaviour, whether it be in normal development of the embryo or in pathological conditions like cancer. Isolating the first protein which could turn a gene on and off would be a major advance and would enable the process to be understood in molecular terms. Evidence for the existence of the controlling protein was at this time only indirect and came from genetic experiments. As he told the science journalist Philip J. Hilts, Ptashne saw it as a great problem:

as I looked into it more . . . it became clear that the others were willing to take risk only to a certain point. The question was, how hard are you willing to work, are you willing to work with the possibility that you'll have nothing at all to show for it? You may work for two or three years, simply fail and look like a fool. If not a fool, at least empty-handed.

Ptashne took that risk.

One floor below Ptashne in the same Harvard laboratory was another molecular biologist, Wally Gilbert, who was also trying to isolate the repressor, but by a different route. (He later won a Nobel Prize for other work on sequencing DNA.) Their research was independent and competitive, but with mutual support and openness. Ptashne worked unbelievably hard, even to illness through exhaustion. 'I think the most important experience you have as an experimental scientist is realizing the extent to which you can be fooled, the extent to which your impulses and aspirations lead you to believe things which have nothing to do with the way things actually work.' As Hilts puts it, the chief experience of science is failure. But within eighteen months Ptashne had isolated the repressor, and so had Gilbert: the honours were shared.

Ptashne's style is still unashamedly aggressive: 'I do needle and goad students, at least those for whom I have the greatest respect. The reason is most people do not understand just how difficult science is, how difficult it is to do something truly first rate or original.' Persistence in the face of failure is a repeated theme in successful science.

The discovery of messenger RNA provides an example of the complex nature of scientific discovery and of a case where illumination was of the '*Eureka!*' type. Proteins are synthesized on small particles in the cell called ribosomes. Messenger RNA is a key molecule carrying the information for proteins from the DNA to the ribosomes – it specifies the sequence of the protein's amino acids (Chapter 1). Ribosomes themselves are made up of protein and another sort of RNA.

Francis Crick has related how it was not at all easy in the late 1950s to get across to other scientists the idea of the genetic code, namely that sequences of DNA were coded for specific amino acids and so provided the code for protein structure. There was a feeling in the larger scientific community that Crick and his colleagues were oversimplifying things. Moreover, they were having great difficulty finding out what the code actually was. They had, however, got the main outlines right. 'But we made one terrible, terrible bloomer. In modern terms we would express it by saying we thought that the ribosomal RNA was the messenger RNA, and that held us up, oh, for several years. The penny dropped one day, one Good Friday, I think it was.' They thought that, because proteins were synthesized on ribosomes which contained RNA, it was ribosomal RNA that carried the code. They experienced a moment of great insight – similar in a way to the discovery of the structure of DNA – when, in a very short time, the whole subject came to look quite different. Only once this step had been taken could there be real progress, and in this case the genetic code was worked out within a few years.

The bloomer came about as follows. By 1957 Crick's 'central dogma' was generally accepted, namely that DNA makes RNA which makes protein. It was also known that proteins were made on ribosomes, which also contain RNA. It was thought that the RNA in the ribosome was the same as the RNA that coded for the protein, but this posed a severe problem for the control of protein synthesis. François Jacob had shown that changes in the amount of synthesis of specific proteins are rapid and under genetic control. Ribosomes, by contrast, are rather stable, and this was inconsistent with the rapid turning on and off of protein synthesis. How, for example, could new ribosomes be made so quickly? Crick and his colleagues were stuck and searched for some way out. Even heretical ideas, such as DNA making protein directly, were considered.

Sydney Brenner was acutely aware of the problem, and on Good Friday 1960 several of the key people, who were in London for a meeting, came to his rooms in King's College, Cambridge. Jacob took them over his experiments on the rapid change in synthesis again; these experiments had been repeated and were now even more persuasive. If the synthesis of a new protein could be rapidly turned on and off, it was hard to reconcile this with it being the gene that controlled this protein if the protein was being made on a ribosome. It may be relevant that the French group, to which Jacob belonged, was more interested in genetic switches than in the problem which occupied the British group, namely the genetic code. In the discussion in Brenner's room, Jacob described an experiment which had been done by some Americans in Berkeley which showed that, for the protein to be synthesized, the gene had to be there all the time. It seemed that the gene needed to be active all the time its protein was being synthesized. This suggested that the gene might be involved by producing an unstable intermediary which would decay and disappear in the absence of the gene. 'That's when', as Crick says, 'the penny dropped and we realized what it was all about.' They then recalled an experiment by some other American workers, who had found a species of RNA that resembled DNA but which they had thought to be some precursor of DNA synthesis. Now Crick and Brenner realized that this RNA was an unstable messenger that carried the information for making the protein from the DNA to the ribosome. It had already been discovered, but the Paris group had not realized it. The ribosome *was* a structure for making proteins, but its RNA was something with a different function and the ribosome required an RNA message from the DNA which specified which protein was to be made. Brenner now saw how they could test the idea of a messenger RNA, and he and Jacob planned the experiment that day. Brenner and Jacob were already going to the California Institute of Technology, and they planned to do the experiment in Mesehlsohn's laboratory there, since he had the right techniques available.

Their new ideas were treated by most Americans with scepticism; the great Delbrück told them, 'I don't believe it.' Their planned experiment was to infect bacteria with a bacterial virus – a phage – which resulted in new protein synthesis for making a new phage, to find out if new ribosomes were made or whether, as they predicted, a new messenger RNA went to the pre-existing ribosomes. The experiment involved density centrifugation to separate out the ribosomes according to whether or not they had incorporated heavy isotopes of carbon and nitrogen. Things went badly wrong. They couldn't get the isotopes into the ribosomes. With only a few days left, they spent the afternoon on the beach. Brenner was uncharacteristically silent but suddenly leapt up shouting 'It's the magnesium.' Running through the experiments in his mind, he had suddenly realized that they hadn't added enough magnesium and thus had damaged the ribosomes. They rushed back to the laboratory and repeated the experiment with the addition of more magnesium – an apparently trivial but crucial component. The experiment worked and the existence of messenger RNA was established. It took six months more of hard work to complete the work in Cambridge.

This discovery is a nice example of sudden insight coming to a group who were making no progress with a problem. Its solution required both imagination and knowledge, and a large infrastructure of work by others. Crick, Brenner and Jacob had an enormous knowledge, in detail, of many, many of the experiments. The trick was to know which experiments were the relevant ones. It may not be easy to find an analogy to this sort of creativity in the arts.

The discovery of messenger RNA is particularly satisfying because the moment of discovery can be pinpointed, the moment of insight recorded. But this is not necessarily typical of progress in science, which is often by slow accumulation such that the breakthrough comes without drama. It is certainly possible to imagine a scenario in which the structure of DNA and the revolution it brought came piecemeal and involved players less charismatic than Crick and Watson; the discovery might then never have acquired its enormous appeal and public exposure. One can see such a case with an equally important advance – the recognition, during the second half of the nineteenth century, that chromosomes were the carriers of heredity. This came by the accumulation of small but crucial advances, but without drama or association with any one scientist.

Thus there is a question about the essential role of genius in science. To what extent are new ideas, and the whole progress of science, really determined by the work of scientists of genuis? The Ortega hypothesis, taken from José Ortega y Gasset's *The Revolt of the Masses*, asserts that genius is not necessary and that 'experimental science has progressed – thanks in great part to the work of men astoundingly mediocre, and even less than mediocre.' Science accommodates and even needs the intellectually commonplace. According to this view, science proceeds, in certain areas at least, by addition of small if not tiny steps, and there are no real breakthroughs.

Some evidence against this idea comes from analysis of the use of the scientific literature. It turns out that 85 per cent of science literature – that is, papers in scientific journals – is quoted in other papers once or not at all each year, while only 1 per cent is quoted five or more times. In the arts and humanities, 98 per cent of papers published are not cited in the following four years, compared to about 40 per cent in science. This supports the argument that an extremely small proportion of the literature is dominant. In cell biology the evidence is similar. About ten key journals dominate the field, but a further 150 journals publish occasional papers which are regarded as being essential. While such key journals may dominate a field, it is far from clear to what extent they rely on the infrastructure created by lesser scientists. The question is less one of breakthroughs than of significant contributions.

The Ortega hypothesis is partly dealing with the issue of whether science proceeds gradually or with sudden jumps: whether progress is slow and gradual, with many contributions, or is due to the work of rare revolutionary scientists. The historian Thomas Kuhn, in his book *The Structure of Scientific Revolutions*, designated as advancers in science those who practise what he calls 'normal science'. They contribute by determining significant facts, by matching facts with theory and by articulation of theory itself, but they remain within a given paradigm – that is, they work within the framework of the dominant ideas current at the time. By contrast, the revolutionary scientists, like Darwin and Einstein, change the paradigm. It has been asked why, if revolutionaries are accorded so much acclaim, everyone does not opt for that mode of science. An answer may be provided by the state of the science – whether, for example, the conditions are right for revolution – but a more likely answer is because it is very, very hard to think of revolutionary ideas.

There are usually lots of other scientists thinking very hard about the central problems, so there are many examples of multiple discoveries. Wallace arrived at evolution by natural selection at about the same time as Darwin. Methods for determining the sequence of bases in DNA – fundamental to genetic engineering – were discovered independently by

Gilbert and Maxam in Harvard and by Sanger in Cambridge. The 'rediscovery' of Mendel's laws of genetics at the end of the last century was made by at least three biologists. The unification of two of the fundamental forces of nature was achieved independently by several physicists (see Chapter 5). And the discovery of the AIDS virus was claimed by both American and French virologists. The list is long.

The traditional interpretation of multiple discoveries is that they show that scientific advance lies outside the individual and rather that the scientific milieu at a particular time – the *Zeitgeist* – determines the nature of the contribution. According to this view, discoveries are inevitable and science does not depend on acts of genius. (This is, of course, a *non sequitur*, for why should there not be several geniuses around at any one time?) There is, in a sense, a certain inevitability of discovery when the appropriate knowledge is available and enough gifted investigators are focusing on the problem.

There is a widely held view – which I believe to be mistaken – that serendipity plays an important role in discovery. This unfortunate word was coined by Horace Walpole in 1754 to describe people's discoveries 'by accident and sagacity, of things they were not in quest of'. I say 'unfortunate', for the word has been rather consistently misapplied to scientific discovery. Again and again one reads reports of accidental or chance discoveries. Examples of serendipity in science are said to abound: Fleming's discovery of penicillin, Becquerel's discovery of radioactivity, the discovery of tranquillizers, and on and on. In each case luck is ascribed a major role in the discovery. But was it really luck, or accident, that was important? May not the whole idea of serendipity be based on a misconception about the nature of science, and also about the nature of chance itself? Even a casual examination of each of the so-called examples of serendipity does, I believe, allow one to reach a quite different conclusion. It will confirm the intense self-awareness that is involved in scientific research: scientific research is based not on chance but on highly focused thoughts.

Louis Pasteur, the outstanding French biologist and doctor, had a reputation for being lucky. At the age of twenty-five, shortly after receiving his medical qualification, he was studying racemic acid, a chemical that is deposited in wine casks during the fermentation of grapes. Pasteur was puzzled by the already established observation that a solution of racemic acid had no effect on a beam of polarized light, whereas tartaric acid, with an apparently identical chemical composition, rotated the beam in a particular direction. So he prepared crystals of racemic acid, and when he examined them under the microscope he noticed that there were two kinds of crystal which, like left and right hands, were mirror images of each other. Distinguished chemists had examined the crystals before but had missed this subtle difference. Pasteur now showed that the right-hand crystals were like tartaric acid, and it was because racemic acid was a mixture of left- and right-hand crystals that it did not rotate the light. This research opened up the whole field of handedness of molecular structures. Life itself is largely built on one class of handed molecules: the left-handed amino acids which are the chemicals from which proteins are made.

The claim for Pasteur being lucky is based on some of the special properties of racemic acid: the particular form he studied is unique in providing crystals which can be recognized under the microscope, and also the separation into the two forms occurs only at temperatures below 26°C. But this is no more luck than that he actually decided to study racemic acid, had a microscope and so on.

Another example of Pasteur's so-called luck was his discovery of immunization using dead bacteria. He was experimenting with the bacteria that caused cholera in chickens. The bacteria were grown on agar plates and were then used to infect the chickens. On one occasion the chickens were inoculated with an old culture in which the bacteria seemed to have died. The chickens did not get the disease. But when these same chickens were later inoculated with a fresh culture they survived the infection. Pasteur, fully aware of Jenner's work on innoculation with cowpox to prevent smallpox, recognized the similarity, and in Jenner's honour called the process vaccination, from the Latin *vacca* (cow), referring to the cowpox work.

When it was suggested to Pasteur that many of his great achievements depended on luck, he replied – I'm sure with more than a little irritation – 'In the field of observation in science, fortune only favours the prepared mind.' It is not by chance that it is always the great scientists who have the luck.

In 1908 Alexander Fleming passed his final medical examination and was awarded the Gold Medal of the University of London. He wrote a thesis on 'Acute Bacterial Infection' for the competition for the Cheadle Medal at his medical school, which he won. In this essay he described what were then thought to be all the defences against bacteria that the medical profession could offer. These were, in addition to the patient's own resistance, the use of antiseptics, some antibacterial agents for specific bacteria (such as mercury for syphillis) and vaccines. Vaccines were the passion of his chief, Almoth Wright, and the treating of bacterial infections was his major preoccupation. Just a year later Fleming had the opportunity to administer Salvarsan, the chemical discovered by Paul Ehrlich which killed the organism that caused syphillis. The astonishing success of this treatment must have made Fleming realize that bacteria could be killed by specific chemical treatment. But, how, he must have wondered, was one to find such chemicals? During the 1914–18 war Fleming was in France, where he observed that antiseptics were powerless to prevent infection of wounds: the results were, unfortunately, often better if no antiseptics at all were used.

If all this is not enough to persuade the reader of Fleming's prepared mind, his discovery of lysozyme must remove all doubts. In 1922 he added a little of the mucous from his nose, at a time when he had a cold, to a bacterial culture. Around the drop, all the bacteria were killed. With great care and patience, he showed that the active component of the mucous which caused this was a natural constituent of tears, which he called lysozyme. This, he thought, was the body's natural protection against bacteria. So, when that fateful mould of *Penicillium* floated on to his bacterial plate in 1928 and he saw that all round the mould the bacteria had been killed, no mind could have been better prepared. His biographer André Maurois reports Fleming as saying:

I had often seen such contamination before. But what I had never seen before was staphylococci undergoing lysis [breakdown] around the contaminating colony. Obviously something extraordinary was happening. With the background that I had, this was much more interesting than my staphylococcal research, so I switched promptly. I am now glad that for years my interest had been directed to antiseptics and that some years before I had found in a somewhat similar manner another naturally occurring antiseptic, lysozyme. But for the previous experience it is likely that I should have thrown the plate away, as many bacteriologists had done before me.

Of course it was fortunate that the mould landed on Fleming's plate, but was that chance any less or more likely than that he had been born at the particular time, had become a doctor and had then chosen microbiology? Thousands, if not millions, of small events shape all our lives. Why focus on just one? To do so is quite misleading. To designate some scientific discoveries as serendipitous can be equally so. If Bobby Jones had discovered penicillin while playing golf, that, perhaps, would have been an example of serendipity.

In 1896 Henri Becquerel was experimenting with uranium salts which emitted light in the dark after being exposed to the sun. He had concluded that the sun had caused the uranium to emit some sort of radiation, because it could blacken a photographic plate. Because, apparently, the sun failed to shine and so delayed further experiments, he developed the photographic plate nevertheless, found that the plate was blackened even though it had not been exposed to the sun and so discovered radioactivity. But this had nothing to do with luck: to develop the plates without exposing the uranium to sunlight was an obvious control that any scientist would be expected to do.

The discovery of the vulcanization of rubber relates more to technology than to science, and indeed many, if not most, of the examples that are supposed to illustrate serendipity are concerned more with the discovery of a process or substance that has useful applications rather than being related to pure science. Charles Goodyear devoted an enormous amount of effort to making rubber impervious to temperature changes – in fact it became an obsession. He treated rubber with a variety of substances, including sulphur, with no success. In

1844, by accident, he allowed a mixture of rubber and sulphur to touch a hot stove and to his surprise he found that it was only slightly charred but, most dramatically, was flexible and tough, over a wide range of temperatures. Vulcanization had been discovered and it seems to me the perfect example of how technology advanced both before and after the Greek discovery of science. No science was required, only curiosity, and the common sense to try a variety of methods and to select the ones that work.

In modern science I am always impressed by the fact that it is always the best scientists who seem to be the luckiest. Of course, advance in science, being a journey into the unknown, must inevitably confront scientists with the unexpected. This is not luck or chance: it is of the very nature of science, for as one explores phenomena or ideas at the frontiers of scientific knowledge it is the unexpected that provides the clues to guide further work. In recognition of this, Bruce Alberts, a distinguished molecular biologist, has cogently argued against giving too many grants to any research scientist, otherwise the scientist no longer has contact with the phenomena and merely receives filtered reports from more junior scientists. Under these circumstances, Alberts rightly observes, important and unexpected observations will escape notice by the leading scientist who has the skill and experience to recognize their significance.

There are, indeed, numerous examples where scientists have, with hindsight, missed the importance of a particular event. In a sense, Aristotle failed miserably in this respect: he should have recognized the contradiction in his ideas about falling bodies, and he certainly could have discovered laws relating to the swing of the pendulum. It shows the absurdity of the idea of serendipity when one realizes that it was nearly 2,000 years after Aristotle before Galileo took notice of the pendulum-like swing of an altar lamp. A more recent example is provided by Peter Medawar. In *Pluto's Republic* he points out how he and his colleagues missed the significance of an observation which, if correctly interpreted, would have led them to recognize a new and very important phenomenon in immunology. This was the reaction of a graft against the host to which it had been transplanted; at the time, they were focusing their attention on the reaction of a host to a graft. We are surrounded all our lives by innumerable 'facts' and 'accidents'. The scientist's skill is to know which are important and how to interpret them.

5
Competition, Cooperation and Commitment

Among the many misconceptions of science are that scientists either pursue truth in a dispassionate manner, their only reward and aim being the better understanding of the world, or that they are entirely competitive and selfish. While both have some elements of truth, these are misleading images. Scientists are emotionally involved in their work, and, in addition to the joys of discovery, the social interactions between scientists play a fundamental role in setting scientists' goals. Scientific knowledge is cumulative, and scientists have a special relationship to other scientists both because they are in competition with them and because they want their esteem, so they cooperate with them. Scientists want other scientists to accept their ideas, but the acceptance of new ideas is more complex than just judgements about verification or falsification. Scientists do not like to give up their ideas or accept those of others without good reasons.

Compared to the creative arts, science is ultimately an anonymous enterprise. Scientists add to the body of scientific knowledge, and it is in essence irrelevant that some are made temporarily famous through a discovery, for in the long run their ideas are incorporated into a common body of public knowledge. For example the invention of the calculus, in the seventeenth century, revolutionized mathematics and is the basis of all modern applied mathematics and engineering. But, other than historians, no one is interested that it was discovered independently by Leibniz and by Newton, who fought bitterly about priority, and no one would now read their almost impenetrable papers. As ideas become incorporated into the body of knowledge, the discoverers, the creators (of whom there may be many), simply disappear. Likewise, no one reads Watson and Crick's original paper if they want to know about DNA, or Darwin if they wish to understand evolution (though it must be admitted that to read *The Origin of Species* can still be very rewarding). Thousands of scientists have contributed to our understanding of DNA and evolution, and their knowledge has been distilled into general and specialized textbooks. Scientific papers have a short life – even really important ones are no longer referred to after a few years. Scientists cannot work in isolation, because the enterprise is essentially cumulative.

Compare all this with the arts: for painters, novelists and poets, the original creation is all-important. The artist does not contribute to a common enterprise; the artist's work is not assimilated into a larger body and its essence is its individuality. For the scientists, by contrast, the aim is to get others to accept their ideas, to obtain consensus. As the mathematician David Hilbert once expressed it, the importance of a scientific work can be measured by the number of previous publications it makes it superfluous to read.

A peculiar feature of science which has important implications for the social behaviour of scientists is that discoveries can be made only once. Once a particular discovery has been made, others cannot make it – though it will, of course, open up new possibilities. The general theory of relativity or evolution by natural selection or the structure of DNA cannot be discovered again. Shakespeare's *Hamlet* was not a discovery: it didn't stop others from writing plays even on related subjects. But Watson and Crick's discovery of the structure of DNA was quite different: once they had discovered the structure, no one else could do the same and a major problem had been solved. Writing *Hamlet* solved no problems in this sense. Knowing the structure of DNA opened up an enormous field of research and there were other discoveries to be made – indeed, several more Nobel Prizes have since been won for work on DNA. Watson and Crick were themselves building on the accumulated knowledge provided by many other workers. And there is another important feature: if Watson and Crick had not discovered the structure of DNA, one can be virtually certain that other scientists would eventually have determined it. With art –

whether painting, music or literature – it is quite different. If Shakespeare had not written *Hamlet*, no other playwright would have done so.

For all these reasons, the strategy that scientists adopt in relation to their work and their colleagues is very different from that of artists. Artists are not subject to the criteria of validation and falsification that are central to the social activity of scientists. Artists may plagiarize, but they cannot falsify in the same sense as scientists can: they cannot cheat.

We are thus confronted with the 'sociobiology' of science. Sociobiology is defined as 'the systematic study of the biological basis of all social behaviour', and sociobiologists ask questions as to why animals engage in the forms of behaviour that are observed. What strategy should scientists adopt to maximize the success of their ideas, which are, in a sense, the scientists' children? Also, how should scientists behave with respect to their subject and their colleagues so as to be most successful? These are questions that sociobiologists ask about animals. An often discussed question, for example, is the basis of altruism. For animals, the answer is framed in terms of the advantage it gives to the survival of the animal's genes. (As the geneticist J. B. S. Haldane perceptively remarked, he would lay down his life if it saved the lives of eight cousins, since that would ensure a better chance for the survival of his genes.) Other questions relate to the investment that animals make in producing and rearing their offspring, which clearly has resonance with scientists' devotion to their ideas. Yet other ideas deal with competitiveness between animals and the extent to which, within a species, aggressiveness pays off. This gave rise to the important concept of an evolutionarily stable strategy: the strategy adopted by animals of the same species, with, say, metaphorical hawklike and dovelike characters competing for the same resource, such that it could not be displaced by a better strategy.

Scientists cannot be treated as idealized animals and it is not legitimate to apply a sociobiological analysis to them. However, it does not seem unreasonable to assume that scientists wish to maximize the success of their ideas. Success can be thought of in terms of selection of their ideas by the community in the field in which they work. This is associated with personal success, which involves advancement in relation to jobs, promotion, praise by one's peers, money for supporting research, some personal financial rewards and, on occasion, prizes. The value to the individual scientist of each of these rewards will vary, but they are closely interlinked and can be lumped together under the rubric of esteem by other scientists.

In order to promote the success of their ideas, and hence themselves, scientists must thus adopt a strategy of both competition and collaboration, of altruism and selfishness. Each must balance his or her behaviour, in relation for example to sharing information, in these terms. Artists are confronted with such choices to a much lesser extent. Another special feature that characterizes modern science is the enormous number of collaborative research projects. Single-author papers are now a rarity in the scientific litera- ture. Many papers have four or five authors, and in some cases in subatomic physics the number of names attached to the paper may be fifty or even more.

It may not be unreasonable to think that the strategy scien- tists adopt is one that is entirely competitive and self-seeking, since there are, in a sense, only a limited number of golden discoveries to be made in any one field at any one time. Once that 'gold' has been claimed, the other 'prospectors' are left penniless. But this view ignores the intensely cooperative nature of the scientific enterprise. Scientific success is not only about making discoveries about nature but about per- suading other scientists of the validity of your ideas. In the process, one has to be part of a community which, with time, has developed quite a rigorous set of unstated norms for acceptable behaviour. Included in these norms are the ideas that science is public knowledge, freely available to all; that there are no privileged sources of scientific knowledge – ideas in science must be judged on their intrinsic merits; and that scientists should take nothing on trust, in the sense that scientific knowledge should be constantly scrutinized. In addition, there have arisen a set of rules for the sharing of materials. In molecular biology, for example, once a paper is published which contains information on specific genes or proteins, then the authors are duty-bound to provide materials from their laboratory which enable other workers to pursue work on those genes or proteins. They may, of course, require that future research be collaborative, but it is not acceptable for them to keep all the materials for themselves.

There is an almost prurient fascination in the media with both competition and fraud in science. It is as if these contaminate the purity of science, and they are viewed almost in the same way as someone of note in the religious world being discovered to be wholly immoral. Competition between scientists is regarded as, at the very least, indecent – quite alien to the image of the ivory-tower scientists pursuing knowledge for its own sake. But this is to fail to understand the special nature of the scientific enterprise and how scientists interact with one another. Scientists have to adopt a special strategy in order to be successful. They have both to compete and cooperate. Carl Djerassi, the chemist who first synthesized the birth-control pill, is one of the very few distinguished scientists who have written a novel about science; it is not surprising that he made fraud and the Nobel Prize its central themes. J. B. S. Haldane is reported to have said that his great pleasure was to see his ideas widely used even though he was not credited with their discovery. That may have been fine for someone as famous and perhaps noble as Haldane, but for most scientists recognition is *the* reward in science.

There are cases where scientists have plagiarized the work of others and where results have been manufactured to support a particular hypothesis. It is inevitable that among the many thousands involved in scientific research there should be a small number who behave dishonestly and quite against the accepted norms. In several cases even distinguished scientists have been involved, by putting their name on a paper containing fraudulent results obtained by a junior colleague. They may, in some instances, not have had the time to examine the primary data in detail and so also have been deceived, since it is one of the dangers of ever-increasing collaborative work that scientists must have complete trust in the colleagues with whom they collaborate. For the functioning and the image of science, fraud is inexcusable; but for the advancement of science in the long run it really does not matter much, because it is so rare. Moreover, many respectable papers will themselves turn out to be wrong or irrelevant. A fraudulent result in an important area will soon be discovered when others fail to replicate the work, and this is exactly what has happened in several cases. More subtle is the scientist's desire to 'massage' the results so as to support a viewpoint. Distinguished scientists have been accused of doing just this. Mendel's results that established his ideas on inheritance were, it is claimed, just too good to be believable. The desire to present one's results in the best light can be difficult to resist. In the example to be given below, Millikan will be seen to have been highly selective about which results he published when measuring the charge on the electron.

Direct evidence for competition in science comes from the finding that 60 per cent of scientists in a survey reported that at least once in their careers some other scientist had preceded them in the publication of their findings. Scientists will be involved in multiple discoveries if they are highly prolific and in an area with many competitors. For this reason, scientists have to take strategic decisions about which areas to work in, since it is a disadvantage to be the 'second' discoverer. The psychologist B. F. Skinner advocates one strategy: 'a first principle not formally recognized by scientific methodologists: when you run into something interesting, drop everything else and study it.' The difficulty is to know what is interesting and when to do the dropping.

The Nobel laureate Barbara McClintock made exactly that sort of decision when, at the age of forty-two and already a scientist of distinction, she made the observation on maize that led eventually to her discovery of what became known as transposition of genes. She came across patches of cells in maize with different colours:

Something had to have occurred at an early mitosis (cell division), to give such a pattern. It was so striking that I dropped everything, without knowing – but I felt sure that I would be able to find out what it was that one cell gained and the other cell lost, because that was what it looked like ... I do not know why, but I knew I would find the answer.

Six years later, in 1950, her talk at a symposium was met with silence and incomprehension. Her ideas were premature. It was very hard to incorporate into current knowledge her idea that pieces of the chromosome DNA moved around, that they were transposed. Stability of the position of genes on a chromosome was fundamental to genetic thinking. It required the work of others on different systems to make her work acceptable and recognized as of fundamental importance. Only in the late 1960s did scientists discover transposition in bacteria. Because of bacteria's very rapid life cycle – minutes, not a year like maize – the system was much more amenable to analysis and could be used to demonstrate the validity of McClintock's theory.

Stories similar to that of McClintock are not all that rare. Two classic examples are Wegener's ideas on continental drift and Lord Kelvin's ideas on the age of the earth. The former was right; the latter was wrong. Briefly, Alfred Wegener, a relatively unknown German geologist, put forward the idea, quite astonishing in the 1920s, that the continents of Africa and South America were once joined together but had, over millions of years, drifted apart. There was enormous opposition to his ideas – even vitriolic hostility. Among the reasons why his arguments were rejected were that they required a major rethink of many geological concepts and that there did not seem to be any mechanism that could provide for the movement of the continents. Only in the 1960s did physicists provide both new evidence, based on measurements of the earth's magnetic field, and a mechanism for movement of the continents which made the proposition acceptable. In a way, the case of Lord Kelvin shows the other side of the coin, for Kelvin was already a very famous physicist and his authority at the end of the nineteenth century was enormous. He opposed ideas suggesting that the age of the earth was much greater than previously thought. He would not accept an age of the order of thousands of millions of years – a time-scale proposed by geologists – because he argued that this was a contradiction of the data available on cooling of the earth. What he did not know, which only later became established, was that the natural phenomenon of radioactivity heated the earth and this rendered his analysis, and objections, untenable. But it took a long time to overcome his opposition.

It should now come as no surprise that psychological studies of the Apollo moon scientists found that those scientists judged the most creative were also the most resistant to change their ideas. All agreed that the notion of the objective, emotionally disinterested scientist is naïve. The image of the disinterested, dispassionate scientist is no less false than that of the mad scientist who is willing to destroy the world for knowledge.

New results that confound current expectations are always treated with suspicion: in fact, it is this critical doubt that determines the way in which a scientific paper is read. The most important features are the title and the summary, for they decide whether one needs to know more. If the conclusions are not surprising, one may not read the results with any great care. If they are novel, however, they will be carefully scrutinized. But if one has reason to doubt the validity of the results one will examine the section on methods in detail.

What, then, determines the acceptance of new ideas? According to Max Planck, 'A new scientific truth does not triumph by convincing its opponents and making them see the light, but rather because its opponents generally die, and a new generation grows up that is familiar with it.' While there are many anecdotes about how, with increasing age, scientists generally become opponents of new ideas, this claim should be treated with caution and not be used to demonstrate the conservatism of science. Scientists do not like to give up ideas to which they have devoted their lives: there is no pleasure in having been wrong. And the resistance to new ideas is not necessarily age-related, for a new theory may simply be wrong. Scientists have to choose the best place for the investment of resources, and so, quite rightly, they will not give up their current theories, even if they involve discrepancies, unless they have something better with which to replace them.

An area of controversy is the claim that there is no rational basis for the objective assessment of rival theories which claim to be able to account for the same set of phenomena:

since the rival theories may use concepts that are quite different, they cannot be meaningfully compared – they are incommensurable. The historian Thomas Kuhn claims, for example, that the concepts of Newtonian and Einsteinian mechanics are so different that they cannot be expressed in the same terms. However, this is disputed by most physicists, who seem to have no difficulty in comparing them, teaching them and showing how Newtonian mechanics can be thought of as a special case of the Einsteinian theory. And where there are conflicting theories in modern science, there are almost invariably ways of devising experiments that would, in principle, distinguish between them.

The idea of incommensurability forms an important part of Kuhn's image of how science works, which he originally set out in his highly influential book *The Structure of Scientific Revolutions* (1962). Kuhn characterizes as 'normal science' those periods when scientists are working within a shared set of ideas which define the field. He terms the dominant conceptual framework the 'paradigm'. 'Paradigm' is a contentious concept, poorly defined, but it nevertheless captures an important component of science. For example, there is a big difference in working within the paradigm of Newtonian mechanics as compared to the Einsteinian paradigm. In Newtonian mechanics, for example, mass and velocity are independent entities; but in Einstein's theory a body's mass changes with its velocity, and space and time are relative rather than absolute. Or, to take two biological examples, there was with Darwin a paradigm-shift away from the constancy of species to an evolutionary paradigm in which species change, and, more recently, the revolution in molecular biology changed the paradigm from metabolism to information. Before the role of DNA was understood, most attention was focused on where the energy for making proteins came from; modern molecular biology introduced the idea that this was not the important issue and that the problem was what information determined the sequence of amino acids in the protein. DNA, as we have seen (Chapter 1), contains the necessary information.

Kuhn has further claimed that paradigm-changes come about through revolutions in science which result from the increasing number of strains put upon the existing paradigm. These strains arise because of the difficulties that are being experienced with the ideas constrained by the current paradigm. Since rival paradigms are regarded by Kuhn as being incommensurable, and so cannot be compared, there is thus no rational basis for the change from one to another: rather, one has to explain the revolutions in terms of the social structure of the scientific community. That is, there is a social process by which the community is persuaded to adopt the new paradigm, since, as we have seen, scientists do not like to give up their hard-won ideas. One may recall Planck's remark that some scientists never do this and the new ideas become established because their opponents die. This may be rather a cynical view. Is it not much more likely that the community will adopt the new view – however painful, like Wegener's ideas about continental drift – when new evidence shows that the new theory provides a more satisfactory explanation? Nevertheless, Kuhn is correct in emphasizing the importance of social process in biology, but in acknowledging this we approach the abyss of relativism (See Chapter 6).

There are indeed examples which show just the opposite to the process claimed by Kuhn. In these cases anomalies – that is, observed facts which are difficult to explain in terms of a current set of ideas – are only recognized *after* a new theory has been generally accepted. Before this, peculiar or uncomfortable evidence may just have been ignored. However, when the new theory appears these anomalies acquire a compelling explanation and are used to support the new concepts. For example, the creationist view in the middle of the nineteenth century held that species were fixed and all animals were made perfectly adapted to their environment. But this was clearly not true of some animals: some ducks with webbed feet did not swim and why should blind animals that lived in caves have eyes? Only with Darwin's theory of evolution by natural selection were these anomalies recognized and explained, and then used to support the theory.

Karl Popper has argued that scientific theories can never be verified, only falsified, and that falsification is the true aim of the scientific endeavour (see Chapter 6). Bold conjecture is to be followed by attempts at falsification. But is this how scientists work? Scientists may pay lip service to the idea of prediction and falsification, but they do not always use it: the process is much more complex. There are, in fact, a number of excellent examples to show this neglect of 'falsifying' evidence. Galileo's comment on Copernicus's theory expresses this aspect forcefully. Copernicus's theory about the movement of planets had difficulties with the phases of Venus, and these difficulties were resolved only with Galileo's telescope, more than fifty years later. Galileo considered it praiseworthy in Copernicus that he had not permitted one unexplained puzzle to worry him. And if Copernicus had indeed known the explanation, 'How much less would this sublime intellect be celebrated among the learned! For, as I said before, we may see that with reason as his guide he resolutely continued to affirm what sensible experience seemed to contradict.'

This neglect of falsification is a stance taken by scientists again and again. Robert Boyle, a giant of English experimental science in the seventeenth century, is an example. Two smooth bodies, such as marble discs, stick to one another when pressed together. Boyle thought that they were held together by air pressure and so predicted that in a vacuum they would come apart. His first experiments did not work, but, rather than give up his hypothesis, he attributed the failure to the vacuum in the apparatus being insufficient. With an improved apparatus he tried again and again until, as described in his experiment number 50, he succeeded:

When the engine was filled and ready to work we shook it so strongly that those that were wont to manage it, concluded that it would not bear to be so much shaken by the operation. Then beginning to pump out the air, we observed the marble to continue joined, until it was so far drawn out, that we began to be diffident whether they would separate; but at the 16th suck . . . the shaking of the engine being almost, if not quite, over, the marbles spontaneously fell asunder, wanting that pressure of air that formerly had held them together.

His conjecture had been shown to be right.

Consider now the famous disagreement around 1910

between Robert A. Millikan in Chicago, and F. Ehrenhaft in Vienna, which has been studied in detail by the physicist and historian Gerald Holton. Their disagreement concerned the value of the smallest electrical charge found in nature – the charge on the electron. Millikan, in his first major paper, pointed out that this value ranks with the velocity of light as a fundamental physical constant. The value of the charge of the electron could be deduced from Faraday's work on electrolysis, but he wished to measure it directly – particularly since Ehrenhaft had reported finding charges of only a fraction of that expected to be carried by the electron.

Millikan's experimental approach was to study the behaviour of oil drops that could be charged such that when a small droplet was moving upwards in an electric field against gravitational pull 'with the smallest speed that it could take on, I could be certain that just one isolated electron was sitting on its back. The whole apparatus then represented a device for catching and essentially seeing an individual electron riding on a drop of oil.' Thus Millikan's technique involved observing single tiny oil droplets in what was effectively a very sensitive balance. In 1910 Millikan put forward a value for e, the charge on the electron, of 4.65×10^{-10} e.s.u. While Ehrenhaft's average value was similar, he also found much smaller values, and in his results the value of the charge seemed to vary continuously.

Holton has examined Millikan's papers and notebooks in detail. In the notebooks used for a 1910 publication, each of the thirty-eight observations is given a more or less personal rating from 'three stars' to none, and the sets of observations are given a weighting from 1 to 7. Millikan was effectively saying that he knew a good run when he saw one. Some observations were discarded altogether because he was unhappy with the experiments. But he goes on to say, 'I would have discarded them had they not agreed with the results of other observations.' In effect he is saying that he has assumed a particular value for the correct results, and that the fundamental charge is a constant. Having examined Millikan's notebooks for the years 1911 and 1912, Holton writes, 'it is clear what Ehrenhaft would have said had he obtained such data or had access to this notebook. Instead of neglecting the second observation, and many others like it in these two notebooks that shared the same fate, he would very likely have used all of these.' The notebooks contain many exclamations such as 'Very low. Something wrong.' 'This is almost exactly *right* and the best one I ever had!!!' 'Agreement poor.'

In the end Millikan's view prevailed and he was awarded the Nobel Prize. He rejected data that did not fit his basic idea, and he would perhaps have justified that in terms of how good the experiment that produced the data was. This is a judgement which all scientists make and which is a crucial feature in distinguishing the good, even great, scientist from the less so. It is that remarkable ability not only to have the right ideas but to judge which information to accept or reject.

Experimental skills themselves should not be under-estimated. Humphry Davy, a great experimentalist in the nineteenth century, recognized how much knowledge was involved in doing an experiment on electricity: 'To describe more minutely all the precautions observed would be tedious to those persons who are accustomed to experiments with voltage apparatus, and unintelligible to others.' And attempts to reproduce some of the experiments of Michael Faraday, an even greater experimentalist, have revealed how much skill is required – and even then it was often difficult actually to see what Faraday recorded that he saw. Indeed, like so many others, Faraday showed considerable determination to continue when he obtained negative results. Even today in molecular biology there are those with 'green fingers'. The ability to get experiments to work is more than just following a rigid set of instructions. If repeating the work of others can be tricky, initiating a new investigation requires even more skill.

It must be admitted that Millikan may have taken his judgement beyond reasonable boundaries; nevertheless, as Holton argues, the graveyard of science is littered with those who did not practise a suspension of disbelief who did hold in abeyance the final judgements concerning the validity of apparent falsifications of promising hypotheses. At least one of the reasons for suspension of disbelief is that experiments are sometimes wrong. One must keep in mind Crick's remark that a theory that fits all the facts is bound to be wrong as some of the facts will be wrong.

There is a relevant story about Charles II, who once invited fellows of the Royal Society to explain to him why a fish when it is dead weighs more than when it was alive. The fellows responded with ingenious explanations, until the King pointed out that what he had told them was just not true.

There are several examples of Holton's principle. The first one illustrates a very important point: falsification can itself be false. There is no guarantee that the experimental falsification itself will not turn out to be flawed. The theory of the physicists Weinberg and Salam on the unification of two of the fundamental forces in matter – the strong and weak nuclear forces in the atom – was tested by experiments carried out in enormous machines – particle accelerators – designed to drive particles to high speed. The initial experiments showed that the theory was wrong. Only later experiments showed that the initial experiments were themselves wrong and the theory was confirmed.

The second example illustrates this point even more clearly, as, unlike with Salam's theory, the experiments were done by the scientist himself.

In 1960 Denis Burkitt, a doctor who had been working in Africa, gave a talk in a London medical school in which he described a tumour, now known as Burkitt's lymphoma, which was the commonest children's tumour in tropical Africa. Not only was this the first description of the disease but Burkitt showed that its causation was dependent on rainfall and temperature. Anthony Epstein, a virologist present at Burkitt's lecture, concluded that the cause had to be a virus, even though the evidence that cancer could be caused by viruses was at that time regarded with deep suspicion and the possibility that human tumours could have a viral origin was regarded as almost absurd.

From that moment Epstein dropped everything else and started working on the tumour. 'Slogging' would be a better description. Material from tumours was flown in from Africa, and he and his group used all the standard procedures for isolation of viruses. All of them failed: the results were, without exception, negative. But he didn't enter Holton's graveyard. He and his team continued the search and tried to grow the tumour cells in culture. Again, failure was complete. Failure continued for over two years; but, although all the laboratory evidence was against the idea that the tumour was caused by a virus, they persisted. 'But it had to be right. It just had the feel of being right. And that's why one carried on.' Then one wintry Friday afternoon a sample arrived from Africa which was cloudy and looked contaminated with bacteria. But Epstein examined it under the microscope and saw that it was cloudy because the tumour had broken up into huge numbers of single cells. Immediately he was reminded of an American group who grew tumours not as lumps, as he had been trying, but by breaking the tumour up into single cells. So he tried to grow the tumour as single cells, and this worked. This was the breakthrough, and soon after he identified a virus growing in the cells – the Epstein–Barr virus had been discovered.

Some final cases come from Einstein's work. Popper has quoted Einstein's statement that 'The general theory of relativity will be untenable if the prediction it made about the gravitational shift of spectral lines were not observed.' But Einstein stuck to his theory even though the prediction was not confirmed during his lifetime. The other example is a very famous case of prediction: Einstein's prediction, again from the general theory of relativity, of the gravitational bending of light. This was confirmed by an English expedition led by Arthur Eddington to observe the eclipse in 1919, and the results from the eclipse created enormous publicity both for relativity and for Einstein. But for Einstein the results seem to have been of much less importance. According to his student Ilsa Rosenthal-Schneider, who was with him when Eddington's cable arrived announcing that measurements had confirmed the theory, Einstein remarked, 'But I knew that the theory is correct.' What, she asked, if the prediction had not been confirmed. 'Then I would have been sorry for the dear Lord – the theory is correct.' This confident view was again expressed later. 'I do not by any means find the chief significance of the general theory of relativity in the fact that it has predicted a few minute observable facts, but rather in the simplicity of its foundation and in its logical consistency.' And Eddington himself even stated that one should not 'put overmuch confidence in the observational results that are put forward until they have been confirmed by theory'.

From examining the history of the field following the confirmation, the historian of science Stephen Brush has concluded that the main value of a successful forecast is favourable publicity: the results from the eclipse put relativity theory much higher on the scientific agenda and provoked other scientists to try to give plausible alternative explanations. But light bending could be considered as reliable evidence for Einstein's theory only when those alternatives failed, and then its contribution was independent of its discovery.

Publicity may seem a strange virtue to ascribe to a scientific experiment but, as we now recognize, that is precisely what scientists need for the survival of their ideas. Science is partly about consensus, and if one's ideas are not widely known they may be neglected. As will be seen, it is social issues of this type that have led sociologists of science to question whether science is anything more than a social construct.

It is unfashionable among historians of science to take what Herbert Butterfield called a Whig view of history – to interpret the past in terms of progress, as opposed to seeing it as a series of events that have no particular direction. But it is precisely in this respect that science, once again, is special: for the history of science *is* one of progress, of increased understanding. Of course there have been errors, innumerable social influences, but, given a reasonable time scale, depending on the subject, progress has been a characteristic of science over the last few centuries. And in the last fifty years the progress in, for example, understanding biology at the molecular level has been astonishing. Science is progressive in that the truth is being approached, closer and closer, but perhaps never attained with certainty. But very close approximation can be a great achievement and is infinitely better than error or ignorance. Philosophers are much involved with such problems.

6

Philosophical Doubts, or
Relativism Rampant

If science is an unnatural process, quite different from ordinary thinking, it might be thought possible to state clearly what the nature of science is and to define scientific method. If only this were so! In fact, defining the nature of science and scientific method with rigour and consistency turns out to be extremely difficult. It is even doubtful that there is a scientific method except in very broad and general terms. Perhaps scientists themselves have helped to create the illusion that method in science is highly ordered, for they write almost all their papers as if there were a scientific method. There is a format of 'Introduction' followed by 'Methods' then 'Results' and finally the 'Discussion'. But, as Peter Medawar pointed out, the scientific paper is a kind of fraud, for its neat format bears no relation to the way in which scientists actually work: imagination, confusion, determination, passion – all the features associated with scientific creativity have been purged from it.

For scientists, defining the nature of science is of only marginal interest, for it has no impact on their day-to-day activities. For philosophers of science, and for some sociologists, by contrast, the nature of science and the validity of scientific knowledge are central problems. These observers have found the nature of science puzzling, and some have even come to doubt whether science is, after all, a special and privileged form of knowledge – 'privileged' in that it provides the most reliable means of understanding how the world works. While providing no real threat to science they have become an increasingly vocal group, with an unfortunate influence on the study of science and its history.

It is the very progress of science that presents the basic

problem. If science provides the best understanding of the world, how should one regard, for example, the ideas about phlogiston that were held before the discovery of oxygen and the understanding of its role in combustion? If those who believed in phlogiston could be so wrong, how can we be sure that the same upheaval will not occur in current areas of science? The whole history of science is filled with new discoveries and the overthrow or modification of ideas which were held to be true. So in what sense, then, is scientific knowledge a true description of the world, and what right have we to call it 'privileged'?

The vast majority of scientists would not be interested in such problems. They would probably just argue that the older theories were the best available at the time, and almost always some, perhaps many, features of an old theory will be incorporated into its successor. Scientists have to accept the possibility that their most strongly held view may turn out to be wrong, but some concepts have been so widely tested that it is extremely unlikely that they will suffer this fate. Even those who are dubious about the privileged nature of science do not direct their criticisms at the results of science itself – that the earth goes round the sun, that water is made of two hydrogen atoms and one of oxygen, or that DNA is the genetic material. The attention of the philosophers, rather, is focused on the nature of scientific knowledge and how it is acquired.

The philosopher Willard Quine, for example, argues – contrary to the experience of scientists – that scientific theories are never logically determined by data, so there are always, in principle, alternative theories that will fit the data more or less adequately. He also argues that any theory can be saved from being falsified by modifying the criteria that are used to decide what counts as a good theory. On this view, widely held by philosophers, any set of empirical observations can always be explained by an infinite number of hypotheses. This view is true only if the hypotheses differ in some very minor manner, like the difference between two numbers at the 100th decimal point. In practice scientists are not concerned with such minute differences except in cases where they will have a real impact on their theories and predictions.

Scientists are concerned not with absolute truth but with theories that provide understanding of the phenomena involved. The criteria for a good theory have already been mentioned (Chapter 1), but it seems that it is up to those who really believe that an infinite number of theories are possible to demonstrate this by providing satisfactory alternatives to classical Newtonian mechanics or to genetic theory. As yet none are forthcoming, and anyone who has tried to construct even a simple quantitative theory to account for some observations will know just how difficult it can be even to get one model to work.

Kuhn's views on incommensurability, with his emphasis on social processes determining the acceptance of a theory (Chapter 5), can lead one to a relativistic view of science. For if there really is no rational way of choosing between rival theories, for choosing between one paradigm or theory and another, then it seems that science may be a mere social construct and that a choice of scientific theories becomes like fashion, a matter of taste. If this were really true then scientific ideas would be merely a reflection of a particular set of social and cultural conditions, and science could not merit the so-called privileged position assigned to it. But such a conclusion is not valid. Although social processes play a role in science, scientists change theories because the new ones provide a better correspondence with reality; because, like Darwin's theory of evolution, they provide a better explanation of the world. While the initial stages of acceptance of one or other of competing theories may have a strong social aspect that involves fashion, power groupings and so on, the main criterion will eventually be how well the theory explains the phenomena.

The emergence of molecular biology is a clear example of a scientific revolution, but not in the way that Kuhn would have us believe. The members of the biological scientific community were not confronted with rival and incommensurable theories between which they found it difficult to choose: rather, scientific advances gave rise to a new set of ideas which completely changed the mode of thought or, in Kuhn's term, the paradigm. Instead of thinking about cells in terms of energy and metabolism, the paradigm changed to information, so that, for example, the questions that were now asked about proteins – key constituents of cell function (Chapter 1) – were not about the source of the energy to make them but about the information for the ordering of the amino acids. Of course there was some resistance to the new ideas and the molecular biologists were evangelical in trying to persuade others. They undoubtedly also used rhetoric. But the evidence from the structure of DNA and other key discoveries was so persuasive that almost everyone – certainly the young – got caught up in the excitement of what is clearly a new age for biology and one which has brought spectacular advances. As the American evolutionary biologist Ernst Mayr has emphasized, it is probably true that philosophers of science have ignored advances in biological science, to their cost. By almost always drawing their examples from physics, they have missed out on revealing examples of scientific progress in other fields, particularly molecular biology.

One of the widely quoted criteria for characterizing science has been Karl Popper's emphasis on falsification rather than verifiability. However, the importance of falsification was also made clear by others like the French biologist Claude Bernard in his book on experimental medicine in 1865. In real life, scientists often do not conform to this formula for doing science, as we have seen in Chapter 5, but there are also some philosophical problems in this approach. It is claimed that verifying a theory is a rather weak way of establishing its validity, and so it becomes difficult to define the conditions under which a scientific theory can be said to be true. Take the trivial hypothesis that all swans are white, or that sodium burns with a yellow flame – 'trivial' because, although they are often used as models for thinking about the 'truth' of scientific ideas, they are not really hypotheses or theories but are just simple correlations from observations and are totally lacking in the richness and explanatory powers of real theories. Popper has argued that the truth or otherwise of these statements cannot be guaranteed on the grounds that they are supported by numerous observations, and so has led the attack on the so-called 'inductive' basis for verification.

If scientists have made thousands of observations that confirm that all swans are white or that sodium burns with a yellow flame, this is, Popper says, no reason to believe that the statement is true. As demonstrated long ago by Hume, induction – inferring relationships from repeated instances – is logically untenable. By contrast, only negative instances – falsifications – provide evidence that can be trusted. If one swan is found that is black, then the hypotheses that all swans are white is falsified definitively. '. . . there is no more rational procedure than the method of . . . conjecture and refutation; of boldly proposing theories; of trying our best to show that these are erroneous; and of accepting them tentatively if our critical efforts are unsuccessful,' says Popper. But would one really give up one's lifelong experience on seeing just one black swan? As described earlier, many scientists would not – and would be unwise to do so – for how could one be sure it was really a black *swan*? Would one not want several examples? If so, one is back with induction. This approach thus avoids the whole question of how scientists actually decide whether or not a theory is refuted or verified. But at least its emphasis on bold conjecture points to a feature of science on which all scientists would agree: science is not just the growth of organized factual knowledge but is a creative endeavour which aims at understanding (Chapter 5). On the negative side, Popper's argument only partly helps define what science is, for, although scientific ideas must be falsifiable, just because ideas are falsifiable does not mean that they are part of science. Absurd ideas are falsifiable but are not part of science, as will be discussed in Chapter 7.

Scientists have an unstated set of criteria for choosing one theory rather than another – and these, moreover, encapsulate some of the main aims of science. In addition to dealing satisfactorily with the phenomena it tries to explain, the theory should have as broad a scope as possible and so encompass a wide range of phenomena. It should be able to predict new relationships and offer scope for further development. It should also be as simple as possible, with a minimum number of hypotheses.

Many of the problems associated with the philosophy of science have their roots in philosophy in general and are not peculiar to science. They are problems relating, for example, to the nature of reality and truth. The existence and nature of ordinary objects such as tables and chairs are held by some philosophers to be problematical. Some philosophers would accept their existence as real, some would deny their real existence, and others would claim that they reflect only external influences on our senses. Thus philosophers are divided among schools of thought whose descriptions – materialism, metaphysical realism, objectivism and so on – hint at their preferred position. But these are the problems of philosophers, and we should not become confused through their inability to deal satisfactorily with the nature of reality and whether or not there is a real world. It seems that Ludwig Wittgenstein may have been saying just this: 'What we find out in philosophy is trivial: it does not teach us new facts, only science does that. But the proper synopsis of the trivialities is enormously difficult and has immense importance. Philosophy is in fact the synopsis of trivialities.'

More generally, if philosophers are correct about the essentially unknowable nature of the world, then this is a problem relevant not just to science but to all knowledge. It must presumably apply to statements like 'The sun always rises in the east' and 'Pigs cannot fly'. For those philosophers who live in a world where they really have doubts about reality, their world is even more unnatural than the world of scientific ideas, but in a quite different way. I have no doubt about the difficulties that philosophers face or the ingenuity they have shown in dealing with such problems. I do, however, strongly deny the relevance of these problems to science. It is essential not to mix up the philosophers' problems in dealing with truth, rationality and reality with the success or otherwise of science. My own position, philosophically, is that of a common-sense realist: I believe there is an external world which I share with others and which can be studied. I know that philosophically my position may be indefensible, but – and this is crucial – holding my position will have made not one iota of difference to the nature of scientific investigation or scientific theories. It is irrelevant.

It is not my intention to argue that science has a claim to absolute validity – indeed, one of the main features of science is that its adherents must be prepared, in principle, to change their minds in the face of evidence. I also must accept that scientists operate within a framework of usually unstated assumptions that the physicist and historian Gerald Holton has called themata. The themata underlie – even underpin – the scientific endeavour and are independent of its subject-matter, experiments and analyses. Copernicus, for example, believed that nature is God's temple and that humans can discern its design and its constant laws – an idea that resonates through Galileo and Newton. Two themata that occur in modern science are the ideas of simplicity and beauty. To these, in physics at least, is coupled the conviction that, as the physicist Steven Weinberg has said, 'we will find the ultimate laws of nature, the few simple general principles which determine why all of nature is the way it is . . .' This echoes Newton, who, after showing how his theory of gravity enabled him to deduce, in detail, the motion of the planets, wrote, 'I wish we could derive the rest of the phenomena of nature by the same kind of reasoning.' And Einstein taught that the noblest aim of science was to grasp the totality of physical facts, leaving out not a single datum of experience. How unnatural, in a way, these themata are: for what in the myriad and varied events of our daily life gives a hint that such a unity – beautiful and simple – might exist?

The physicist John Barrow has listed a further set of assumptions:

- There is an external world separable from our perception.
- The world is rational: 'A' is not equal to 'not A'.
- The world can be analysed locally – that is, one can examine a process without having to take into account all the events occurring elsewhere.
- There are regularities in nature.
- The world can be described by mathematics.
- These presuppositions are universal.

These assumptions may not be philosophically acceptable, but they are experimentally testable and they are consistent with the ability of science to describe and explain a very large number of phenomena.

Has philosophy in fact influenced science? Many of the leading physicists at the beginning of the century were well schooled in philosophy, and the German physicist Ernst

Mach had strong views on the nature of science. However, an interest in philosophy was just part of the 'normal' intellectual cultural environment in Germany at that time. Today, things are quite different, and the 'stars' of modern science are more likely to have been brought up on science fiction. They view the philosophy of science as, in Holton's phrase, a 'debilitating befuddlement', and it has been remarked that the physicist who is a quantum mechanic has no more knowledge of philosophy than the average car mechanic. Not only are most scientists ignorant of philosophical issues, but science has been totally immune to philosophical doubts. In this century at least, science has generally been wholly unaffected by the philosophers of science, though some Nobel laureates, like the neurophysiologist John Eccles, claim that their work has been greatly influenced by Popper. Another possible exception is psychology, where there is a link because psychology is closely related to problems that have origins in philosophy, such as the nature of knowledge and how the brain thinks.

Even distinguished philosophers of science like Hilary Putnam recognize the failure of philosophy to help understand the nature of science. They have not discovered a scientific method that provides a formula or prescriptions for how to make discoveries. But many famous scientists have given advice: try many things; do what makes your heart leap; think big; dare to explore where there is no light; challenge expectation; *cherchez le paradox;* be sloppy so that something unexpected happens, but not so sloppy that you can't tell what happened; turn it on its head; never try to solve a problem until you can guess the answer; precision encourages the imagination; seek simplicity; seek beauty . . . One could do no better than to try them all. No one method, no paradigm, will capture the process of science. There is no such thing as *the* scientific method.

Just because it may be difficult to define exactly what is meant by 'life', for example, and thus whether or not certain machines or computer programs are or are not alive, that in no way means that there is no distinction between living and inanimate systems. Science is a complex social process, and no simple-minded description in terms of Kuhn's paradigms or Popper's falsification will provide an adequate description. The demarcation problem is real only in the sense that science is rich, varied, heterogeneous and complex. Its edges may be blurred, but the core is solid.

For reasons that may be connected with the peculiar nature of science, we have a situation in which the ideas of Karl Popper and Thomas Kuhn seem to be far better known among non-scientists, especially those working in the humanities, than the ideas of almost any contemporary scientist. Another widely quoted philosopher of science is Paul Feyerabend, who, in his book *Against Method*, urges his readers to 'free society from the strangling hold of an ideologically petrified science just as our ancestors freed *us* from the stranglehold of the One True Religon'. All these ideas would not matter if they remained in the philosophical domain, but unfortunately they are sometimes used to undermine the scientific enterprise itself, on the grounds that if science's attitudes towards truth and the role of evidence are philosophically untenable then the whole of science is also suspect.

A less philosophical and more pragmatic approach to understanding the nature of science is to examine how scientists do their work. It would be of great interest to know more about the social interactions between scientists and to see how these interactions, and also interactions with other parts of society, influence how scientists work. Scientists do not work in a cultural or social vacuum. In Chapter 5 I raised briefly the issue of the 'sociobiology' of science, and for a formal analysis, one might look to those sociologists who study science.

The more traditional sociologists of science, as represented by the work of Robert K. Merton, sought to understand social processes in science and tried to define, for example, the procedures that most scientists accept and adopt in their work. Older sociologists like Emile Durkheim even excluded science from the sociology of knowledge, on the grounds that it was a special case. I am a great admirer of the sociologist Max Weber, and it is reassuring to know his attitude to science and what he thought it meant: 'It means the knowledge or belief that if one but wished one could learn it at any time. Hence, it means that principally there are no mysterious incalculable forces that come into play, but rather that one can, in principle, master all things by calculation.' Weber recognized science's explanatory power: one no longer needs 'recourse to magical means'. He recognized both the power of the rational experiment and that science does make presuppositions, like accepting the rules of logic.

More recently, however, some sociologists have identified themselves with what is called the Strong Programme of the Sociology of Science. This approach takes the view that the very nature of belief and rationality in science requires expla-nation and the same sort of analysis as non-rationality. No distinction appears to be drawn between good and successful science and what most scientists would regard as second- or third-rate work. Those who hold to the Strong Programme believe that all knowledge is essentially a social construct and so all science merits the same attention. All knowledge is regarded as relative to the social environment in which it is constructed. This new-style sociology, which claims the relativity of science, is called the sociology of scientific knowledge and is known by the acronym SSK.

The SSK approach to science is as follows. It feels bound to ask whether a belief is part of the routine cognitive and technical competence handed down from generation to gener-ation and is supported by the authorities of the society. Is it transmitted by established institutions or supported by accepted agencies involved in social control? Is it bound up with patterns of vested interest? Does it have a role in further-ing shared goals, whether political or technical or both? What are the practical and immediate consequences of particular judgements that are made with respect to the belief? The most striking feature of this approach is that it says nothing about the belief's contribution to understanding, its corres-pondence with reality or its internal logical consistency.

The SSK programme on one central point is explicit: 'we should abandon the idea of science as a privileged or even a separate domain of activity and enquiry.' For such sociol-ogists of science, like Steve Woolgar, the certainties about science – that is, the old beliefs in its cultural uniqueness – have gone. Relativism is strongly defended by proponents such as Barry Barnes and David Bloor, who claim that the real threat to a scientific understanding of knowledge and cognition is posed by those who oppose relativism and who grant certain forms of knowledge, such as science, a privileged status. They hold strongly to what they call their equivalence postulate, which is that all beliefs are on a par with one another with respect to the causes of their credibility. For them the incidence of all beliefs, without exception, calls for empirical investigation, and beliefs must be accounted for by finding the specific local causes for their credibility. Such strong statements make one wonder whether they accept the reality of everyday items like cups of tea.

Even statements that $2 + 2 = 4$ are treated as legitimate targets for sociological questioning, and so too are logic and rationality. It is claimed that, 'By looking at reason and logic, we find that reason, logic and rules are post-hoc rationalizations of scientific and mathematical practices, not their determining force.' In other words, SSK is an extremely ambitious programme with very large, not to say extravagant, claims. It is thus necessary to examine some of the evidence on which the claims are based and what new insights this approach has given us. One can say at once that there have been very few SSK studies of mathematics and logic, and it is not unreasonable to ignore SSK claims to success in those disciplines. But in biology and physics there have at least been a number of studies, and I will describe some of these.

When they discover a new law, a new phenomenon, a new object, scientists believe that the discovery relates to an existing external world. To count as a discovery, their findings should be novel and preferably important. A very different view is taken by those who adhere to the SSK programme, who wish to emphasize that it is the social content that determines whether or not something is called a discovery. Instead of asking about the characteristics of scientific discovery, the newer perspective tends to ask: given that scientists' actions and beliefs can be organized in various ways, by what interpretive practices is science made to exemplify a certain kind of rationality?

The discovery by Mendel of the fundamental laws of genetics has been examined by Augustine Brannigan within this sociological framework. Contrary to the widely held view that Mendel's paper of 1866 was ignored until it was rediscovered in 1900, Brannigan argues that it was less the content of Mendel's paper than the context within which it appeared that led to it being hailed as a discovery in 1900. That context, Brannigan wishes to emphasize, was related both to a priority dispute between the geneticists Carl Correns and Hugo de Vries and to disputes about its relevance to evolution. There is, in fact, evidence that Mendel's paper was not completely neglected when it was first published but was quoted a number of times, though no one gave it any prominence or suggested that it was significant. There is some uncertainty as to where and when de Vries, who arrived at laws similar to Mendel's, first came across Mendel's work. Whatever the case, when Correns, who had also discovered similar laws, received a reprint of de Vries's article on his discoveries, on 21 April 1900, he at once sent off a paper in which he announced results similar to those of de Vries but gave priority to Mendel. It is not unreasonable to suggest that he did this in order to resolve a priority dispute. It is also reasonable to see Bateson's vigorous support for Mendel reflecting in part his belief that Mendel's results supported his views that discontinuous variation was the key feature in evolution. Thus, in Brannigan's view, Mendel's fame is due less to his science than to how this was used by others to promote their own positions.

To be identified with a discovery is prestigious in science and is a major reward for a scientist. We should not be surprised that Correns may have used Mendel's work in order to prevent de Vries being given priority. Priority disputes are indeed a common feature of science, and the assigning of credit certainly reveals complex social interactions. Merton has described what he calls the Matthew Effect – scientists who are already eminent get a disproportionate amount of credit, at the expense of those less well known. This may be due to a scientist's reputation making a discovery more visible and giving it respectability. The sociologists are correct in claiming that the success or failure of a scientific idea may, initially at least, be due to much more than 'pure' scientific criteria.

Mendel's discovery, later confirmed by de Vries and Correns, showed that inheritance of characters could be understood in terms of the transmission of discrete characters that maintained their identity from generation to generation. These 'discrete characters' were only later to be identified as genes. A major feature of Mendel's work was to allow the study of inheritance to be expressed mathematically and to make it possible to state laws as to how a character would be inherited in subsequent generations. From a scientist's viewpoint, Mendel's approach was new and fundamental. As the molecular biologist François Jacob points out, Mendel's achievement was similar in its power to the introduction of statistical mechanics into physics: he concentrated on a small number of characters with sufficiently striking differences for discontinuity to be introduced. With Mendel, biological phenomena acquired the rigour of mathematics. This was not by accident, for in the introduction to his paper Mendel says 'so far, no generally applicable law governing the formation and development of hybrids has been successfully formulated' and he points to the difficulty of the task. He then speaks of his study in terms of the large number of experiments required as well as the necessity to arrange the different forms 'with certainty according to their separate generations'. He very carefully chose his experimental material with this in mind. The distinguished geneticist R. A. Fisher some time ago remarked that people find in Mendel's paper whatever they are looking for. Yet Brannigan argues that Mendel was not ahead of his time and his reputation was modest because his identity with his contemporaries was so complete.

What the sociologists do not illuminate is why no one had done Mendel's classic experiment before. It is a similar problem to that posed by the enormous gap between Aristotle and Galileo with respect to thinking about motion. There is no doubt that social factors must play a role, but, unless they believe that Mendel and Galileo made significant discoveries, it is unlikely that sociologists will analyse them with respect to this question. If they treat all science in so dispassionate and detached a manner as not to single out great achievements, they may be missing the core of the problem. Scientific discovery cannot be judged only in social terms but must also take into account the new understanding or knowledge it provides.

Another example of historical analysis relates to phrenology – the so-called science of interpreting brain function and capability from the size and shape of the head. Phrenology began with a Viennese, Dr Franz Gall, who, with his assistant Johann Spurzheim, in the late eighteenth century put forward these three main principles: the brain is the organ of mind; it is made up of a number of separate organs, each related to a distinct mental faculty; and the size of each organ is a measure of the power of its associated faculty. Thirty-six faculties were listed, including love of children, blandness, prudence and dignity. In Edinburgh at the beginning of the nineteenth century, an enthusiastic disciple of Spurzheim was George Combe, an eclectic scholar with no formal scientific training. Opposition to phrenology in Edinburgh came both from the anatomists at the university and from those who taught the philosophy of the human mind, particularly Professor Sir William Hamilton. The Scottish moral philosophers held the view that mind is an immaterial entity which is both single and indivisible, and this contrasted sharply with the view of the phrenologists. The philosophers were critical of the thirty-six faculties, and asked why there was not among them, for example, a special faculty for love of horses. (However, that was perhaps not quite fair, for there was a region associated with the love of animals.) The phrenologists and those in the university were in vigorous dispute between 1803 and 1828, when the case for the phrenologists collapsed because of evidence relating to observations on the brain. A crucial issue was the frontal sinuses, since, according to Hamilton, they vary greatly and mask the development of about one-third of the so-called phrenological organs. Another point of criticism was that neither the phrenological organs nor the associated faculties were clearly defined – almost any observation could be confirmed. Hamilton also found that the size of the cerebellum – the supposed organ of sexual activity – was larger in females, completely contrary to the phrenologists' expectation.

However, this quite conventional view that scientific evidence resolved the issue has been criticized because of its neglect of the social dimension of the controversy. It has been suggested that the viewpoints of the opposing factions were incommensurable, and that the war between the phrenologists and the moral philosophers should, in large measure, be treated as a conflict between university professors and those exposed to university teaching, on the one hand, and those not associated with the university, on the other. There was, for example, considerable support for phrenology from the lower, middle and working classes. The phrenologists' emphasis upon empirical methods in mental science reflected their socially based anti-elitism, and their great claim was that phrenology was founded on observations which anyone could make and so enabled the ordinary man to discover the truth. They thought, mistakenly, that science is a common-sense activity.

The sociological viewpoint does not regard the difference in the intellectual bases of the two sides as important, even though untrained amateurs have rarely made a significant contribution to any important scientific controversy. Moreover, the conflict was not really within science but between those in science and those outside science. It seems that the SSK analysis is based on the belief that anyone can do science and that the difference in training would not be relevant. Of course, in the end, it is only the fit with reality that matters.

It is curious what topics the SSK have chosen, or not chosen, in order to study relativism in its historical context. Hardly any of the major achievements of, for example, modern biology – the gene and DNA, electrophysiology, biochemistry and so on – has been the subject of an analysis which supports the relativists' case; and among the examples of creativity given earlier it is not easy to see how the discovery of messenger RNA (Chapter 4) or the structure and role of DNA (Chapter 1) could merely be social constructs. They could only appear to be so to someone ignorant of the complex science involved.

One of the rare examples of an attempt to analyse a major advance in modern science is Andrew Pickering's study of the revolution in high-energy physics that resulted in the discovery of quarks and the unification of two of the fundamental forces of nature – the weak nuclear force and the electromagnetic force. Quarks were seen as a new fundamental entity from which many of the particles in the world of the subatomic physicist were built. A new era of physics emerged. Pickering's claim is that he has written his history – and it seems that his historical account is excellent – within the framework of the 'constructivist' approach to the sociology of science. He is indeed an adherent to the 'Strong Programme' of the sociology of knowledge. However, there is really nothing in his analysis that reflects such an approach or that is in conflict with an image of scientific advance that scientists themselves would readily find acceptable. In fact there is little sociological analysis at all where one might expect it – of, for example, the factors involved in the financing and building of the very expensive high-energy accelerators that were required for the experiments. His account shows just the sort of complex interactions between theoreticians and experimentalists that one might expect. On several occasions the experiments failed to support the theory (which was eventually to be successful) and were later set aside, even though they were never discredited. (This is, of course, another counter-example of Popper's falsification model being the way in which science advances.) What Pickering does make clear is the symbiotic relationship between theoreticians and experimentalists: both are looking for new opportunities to advance their work.

A central point for Pickering is that scientific choice is in principle unlimited and open, for he believes that the choice of a theory is not determined by any finite set of data. He takes the view that it is always possible to invent an unlimited set of theories to explain a given set of facts. But, as we have seen, this is a quite misleading view. The puzzle lies in how scientists decide which experimental data or which theoretical construct they are willing to give up when these are in conflict. Pickering's claim is that such decisions are not forced on the scientists by the data but represent an open choice, in which options are foreclosed according to the opportunities that are perceived for future practice. But this is merely a claim: it is not demonstrated. Just the opposite is demonstrated by his own reconstruction of a crucial development, the acceptance of the neutral current – a phenomenon associated with particles known as neutrinos. Before 1971 there was general agreement that the neutral current did not exist, yet from 1974 onwards the neutral current was accepted. In the earlier period the experimental work had given indications of the neutral current, but the calculations involved were filled with uncertainties and, in the absence of an appropriate theory, the issue was not pursued. Only later, when there was new theoretical work which encouraged both re-examination of the previous experiments and initiation of new ones, was the neutral current accepted. If this is to be regarded as the social construction of knowledge then it is quite acceptable to most scientists and is quite uncontroversial. Scientists are dismissed by Pickering for their naïve realism, but he offers nothing in its place. Scientists can be very proud to be naïve realists.

A striking feature of the revolution in high-energy physics relates to Kuhn's concept of incommensurability. The new and old physics are very different and in a sense incompatible. But contrary to the notion of incommensurability, the new physics arose in a congenial world where the issues were widely debated, characterized by mutual congratulation, with little conflict or recrimination, and the new physics helped unify areas of physics that had previously been somewhat isolated from one another.

In attacking the claims of the relativists, I am not arguing that social factors have no influence on science. Quite the contrary: scientific thinking is influenced by the ideas current at the time and often takes concepts from the prevailing technology. Creativity is influenced by many factors. There is no denying that authority, fashion, conservatism, and personal prestige play important roles, and it is no surprise that scientists use rhetoric to promote their ideas – their ideas are, after all, precious to them, and they wish to see them succeed. There is no doubt that, in the 1950s, Francis Crick and other molecular biologists were rhetorical – even evangelical – about their new subject and its new approach. Again, Galileo certainly used rhetoric to persuade and attack. But it is misleading to think, as some have claimed, that science is really nothing but rhetoric, persuasion and the pursuit of power.

No amount of rhetoric is enough to persuade others of the validity of a new idea, but it can make them take it seriously – that is, follow it up and test it. But persuasion ultimately counts for nothing if the theory does not measure up to the required correspondence with nature. If it does not conform with the evidence, if it is not internally consistent, if it does not provide an adequate explanation, the authority and all the other social factors count for nothing: it will fail. Such a failure is undoubtedly culturally determined, the culture being one that adopts a scientific approach.

The case of continental drift (Chapter 5) provides a good example of how new ideas became accepted because of evidence, not social factors, though social factors did delay acceptance. Another example, from the 1960s and early 1970s, is the claims, based on experiments, for the discovery of a new form of water, called polywater, which some even thought could be as dangerous as 'ice-9' in Kurt Vonnegut's *Cat's Cradle* because it could cause all the world's water to be crystallized in a chain reaction and so lead to a catastrophic drought. Although many distinguished physicists were involved, there was, as with continental drift, tremendous scepticism. In this case the scepticism turned out to be justified for the experimental evidence was due to impurities from the glass affecting the water. Again it was evidence, not social forces, that caused the rejection of the idea. And the same scientific process led to the rejection of the recent exciting claims for cold fusion. Small errors in science may go undetected, but this is not the case with major issues: the community can respond vigorously – the institution of science is robust.

Too much emphasis has probably been placed on theory as compared to experiment in considering the nature of science. Experiments are not used just to test predictions of theories: observations are used in a complex manner and involve subtle interactions between the experimental set-up and the observer. There is also a social aspect – the necessity to enable others to do the experiment, and even to use particular experiments to persuade others to one's point of view. A detailed analysis of Faraday's notebooks provides insights into how nature – reality – does influence scientific thinking

and gives the lie to social construction. Faraday had continually to reconstruct, refine and elaborate his ideas as the apparatus provided observations which were quite unexpected. His
problem was to make sense of all the phenomena.

Support for the relativism of science has come from anthropology as well as from sociology, however. It results from
an unwillingness among some anthropologists to regard
thinking in primitive societies as somehow inferior to that
which characterizes the West – namely scientific thinking and
a passion for rationality – an unwillingness to admit that
there is 'little chemistry and less calculus in Tikopia or Tim-
buctoo' and to draw from this the conclusion that science is
not, after all, a common property of mankind. Thus there
have been extensive attempts to show that thinking in primitive societies is rational and is not intrinsically different from
scientific thought (see Chapter 3). African religious systems
are seen as theoretical models akin to science and conflicts
between rational and mystical, reality and fantasy, empirical
and non-empirical are minimized. Such cosmologies may be
internally consistent, given their assumptions, but even that
is not clear. A favoured model is that of Evans-Pritchard's
study of magic among the Azande in the 1930s. A Zande,
according to Evans-Pritchard, 'cannot think his thought is
wrong . . . A Zande cannot get out of its meshes because it
is the only world he knows' and the system is thus rational
and consistent within these constraints. But the system is
closed, compared to the openness of science, and there is no
critical tradition, no alternatives, no confession of ignorance.
There are good reasons for this – isolation, lack of writing –
but even so the system is very different from science. The
anthropological explanations of cosmologies which reflect the
structure of society are very different from the scientists'
cosmology, which tries to explain the universe without reference to human beings.

Unlike science, witchcraft and magic can have a beneficial
effect on community life since magic can be used against
anxiety and social pressures. More generally, 'everything
works' – each part of the system of beliefs contributes to the
maintenance of the whole, no engines idling. The work that
the thought does is social. Nothing could be more different

from science, which has, on occasion, totally undermined the whole conceptual framework – like Darwin's ideas on evolution – and is so often counter-intuitive. There is nothing counter-intuitive for members of a primitive culture, or even a religious one. One should be surprised that neither the sociologists nor the anthropologists have shown much interest in these fundamental differences.

Edmund Leach points out that, taken by itself, myth appears as pure fantasy, but that the way of life of the people who use it is, in fact, ratified by the myth itself. He wishes to dispel the notion that such societies have what earlier anthropologists regarded as childish superstitions, and he suggests that primitive thought is 'just as sophisticated as we are, it is simply that they use a different system of notation.' For Lévi-Strauss, analysis of myth leads him to a surprising conclusion: 'If our interpretation is correct we are led toward a completely different view – namely that the kind of logic in mythical thought is as rigorous as that of modern science, and that the difference lies, not in the quality of the intellectual process, but in the nature of the things to which it is applied . . . we may be able to show that the same logical processes operate in myth as in science.' But, however clever or logical other societies may be, one should not confuse 'logic' with science. And while it must be recognized that science is influenced by and borrows from the society in which it works – for example, ideas about how the embryo developed were strongly influenced by religious beliefs (Chapter 7) – and scientific ideas may, on occasion, be used to justify social attitudes (Chapter 8), what is more important is that the structure and nature of scientific theories from quantum mechanics to the genetic code have been arrived at in quite a different way to the myths studied by anthropologists and serve a quite different purpose: namely, to provide an explanation. 'Understanding' is a word seldom used by anthropologists or sociologists of science.

The issues with respect to both relativism and the importance of sociological influences on science might be encapsulated by asking if one could have had a different science if historical conditions had differed. Would a physics have evolved that is not based on what we now consider to be a set of basic forces? Would a biology not based on cells and DNA have been possible? Would the periodic table or carbon chemisty never have emerged? To the relativists the answer must presumably be 'yes', but then the onus is on them to demonstrate the validity of their position. To me the answer is an unequivocal 'no': the course of science would have been very different, but the ideas would have ended up the same. In my view science, despite blips and errors, more and more provides an understanding of the world. There is one argument that may be persuasive – the role of mathematics.

The quantitative aspect of science is fundamental. Probably even the most ardent relativists do not believe that mathematics is a social construct. Yet some parts of mathematics – often from unexpected areas – provide essential tools for describing particular phenomena. One cannot imagine a science of motion, a successful science, that does not rely on the calculus. If the relativists wish to persuade us of social constructs, they will have to provide, at the least, major counter-examples.

Those sociologists who support relativism have a further problem. They claim that no body of knowledge, nor any part of one, can capture, or at least can be known to capture, *the* basic pattern or structure inherent in some aspect of the natural world; that no particular ordering is intrinsically preferable to all others; and that specific orderings are constructed, not revealed, are invented rather than discovered. These dictates must also apply to their own ideas, which must then themselves be just another invention of no special validity.

Of course the nature of sociology itself provides an obvious, and sociological, explanation for sociologists' wishing to undermine science. In a sense, all science aspires to be like physics, and physics aspires to be like mathematics. But too great an aspiration can lead to frustration. In spite of recent successes, biology has a long way to go when measured against physics or chemistry. But sociology? Biologists can still be full of hope and are going through exciting times, but what hope is there for sociology acquiring a physics-like lustre? One has to recognize that the problems that sociologists are dealing with are enormously complex and at this stage it is premature to expect much progress (Chapter 7). The situation is as bad as, or even worse than, in psychoanalysis. It is thus not surprising that, as Howard Newby, chairman of the Economic and Social Research Council, put it, because of their 'massive inferiority complex' social scientists have 'descended with glee on those who have successfully demystified the official credo of science and who have sought to demonstrate that science is but *one* means of creating knowledge'. For them it then becomes quite unnecessary to have to try to emulate traditional science.

By ignoring the achievements of science, by ignoring whether a theory is right or wrong, by denying progress, the sociologists have missed the core of the scientific enterprise. Science has been extraordinarily successful in describing the world and in understanding it. There is a real need for sociologists to try to illuminate this unnatural process. What is required is an analysis of, for example, what institutional structures most favour scientific advance, what determines choice of science as a career, how science should best be funded, how interdisciplinary studies can be encouraged. Philosophical attacks on science may be healthy in the sense that one should always maintain a critical stance; thus far, however, the results have been disappointing. I must side with Francis Bacon, who, 400 years ago, urged that those interested in science ought to 'throw aside all thought of philosophy or at least to expect but little and poor fruit from it'.

An encouraging and rather novel perspective has been adopted by the philosopher Richard Rorty. Rationality can be taken to mean a way of proceeding which is sane and reasonable, and in which discussion is possible and dogmatism is avoided. It is within such a context of rationality that Rorty sees science as exemplary, since it is a model of human solidarity. The institutions and practices of science can provide suggestions as to how the rest of our culture might organize itself. Leaving aside the question of whether scientists are more objective, rational, logical and so forth, scientists have developed a procedure in which there are free discussion, accepted standards of behaviour and a means of ensuring that truth will, in the long run, win. Truth will win in the sense that open discussion and observing nature constitute the best way of making progress.

7
Non-Science

Pp. 124-150

If scientific knowledge is special and privileged – in the sense that it provides our best understanding of the world – how can we distinguish between science and non-science? How does one deal with claims to be included within this framework of privileged knowledge from those whom most scientists would wish to exclude? There is a continual plea for recognition from those who believe in paranormal events and astrology, and there are more serious claims for recognition from some of those who work in complex areas of human behaviour such as psychoanalysis. There are also the issues of the compatibility of science with religious belief and the claims of, for example, the creationists.

It is not always easy to give good reasons for distinguishing between science and non-science – for dismissing, for example, some claims for paranormal events. Popper's falsification criterion – if an hypothesis cannot be falsified it cannot qualify as science – is unfortunately of little help, since many falsifiable ideas like 'Eating hamburgers will make you a good poet' are just absurd. Falsifiability is a necessary but not a sufficient criterion. For a subject to qualify as science it needs at least to satisfy a number of criteria: the phenomena it deals with should be capable of confirmation by independent observers; its ideas should be self-consistent; the explanations it offers should be capable of being linked with other branches of science; a small number of laws or mechanisms should be able to explain a wide variety of apparently more complex phenomena; and, ideally, it should be quantitative and its theories expressible by mathematics.

For example, there is a question as to whether the social sciences really are science. They can certainly use some of the methods used in the so-called 'hard' sciences – from physics through to biology. Hypotheses can be framed and tested as well as possible. But the problem lies in 'as well as possible': the peculiarity of the social sciences is the complexity of the subject-matter, and so the difficulty of disentangling causal relationships is immense. There is little possibility, for example, of doing experiments equivalent to those in physics, say, in which it is characteristic to try to vary just one variable at a time, keeping others constant, and so observe its effect on the system. A simple case is varying the temperature of a gas while keeping the volume constant and seeing the effect on the pressure. It is this 'isolationist' approach that has been so successful in the 'hard' sciences. Even where correlations between different events are collected by social scientists, it is extremely difficult to provide controls, which are essential wherever real confidence is to be placed in the results. That is why random clinical trials are essential for assessing medical treatments: the effect of a treatment can be judged only by comparing two groups, chosen at random, only one of which is treated. Again, compared to biology, say, it is very hard to be reductionist in the social sciences. The ability to account for much of physiology and anatomy in terms of cellular behaviour, and then in turn to be able to explain cellular behaviour in molecular terms, as yet, has no effective equivalent in the social and psychological sciences.

In order to focus on a particular example of a theory in the social sciences, I have chosen psychoanalytic theory. Freud has provided us with a set of seductive ideas that have had a major impact on how we try to explain human behaviour. I will consider whether or not these ideas constitute a science. I shall try to show that if psychoanalysis is a science at all, it is a very primitive science and, in an important sense, premature. An analogy can usefully be drawn between embryology in the eighteenth century and psychoanalysis today. The problem of whether embryos are preformed or whether form emerges during development can provide a useful model for a science at a primitive or even premature stage.

The first clear-cut body of embryological knowledge is associated with Hippocrates, in the fifth century BC, who viewed development in terms of what he considered to be the two main constituents of all natural bodies: fire and water. This may now seem no less absurd than Thales's passion for water (Chapter 3), but at least it was an attempt to explain the nature of things in terms of a general theory. But the so-called triumph of Greek embryological thought belongs to Aristotle, who believed that the embryo was made out of menstrual blood and that the male dynamic element gave it shape. Aristotle asked whether all the parts of the embryo come into existence together, or are they formed in succession, like the knitting of a net? He thus defined the preformation/epigenesis debate which was to continue for 2,000 years. Having opened chickens' eggs at different times, he argued in favour of the knitting analogy and thus for epigenesis – that is, the gradual generation of embryonic structures. But his rejection of preformation – that everything was preformed in miniature from the beginning – was based on philosophical arguments, not on observation.

Aristotle's theory of development based on epigenesis as distinct from preformation was not based on any real evidence. Though we now know that embryos do develop by epigenesis, he was correct for the wrong reasons: it was little more than an inspired guess. However, he posed important questions about the nature of development, and his influence was enormous. So there was little progress in thinking about development until the late nineteenth century. Fabricius and William Harvey, in the seventeenth century, while providing detailed observations on developing embryos, were essentially Aristotelians, using their observations, with little justification, to support epigenesis.

The notion of preformation in contrast to epigenesis, namely that all embryos had existed from the beginning of the world, was first formulated in detail in the 1670s by, among others, the biologists Malebranche in France and Swammerdam in the Netherlands. Preformation was based on the concept of the first embryo containing all future embryos. There were grave doubts as to whether epigenesis was possible, and it also seemed to undermine God's powers. Could it really be possible for the marvellous development of embryos to be explained by mechanisms based on scientific principles such as physical forces? Preformationists believed, rather, that all the organs were present from the very beginning of development and merely grew larger. If preformation was correct, then the embryo must contain, in miniature, all the future animals (or plants) to which it would give rise. It appeared to Malebranche as not unreasonable that there are an infinite number of trees in a single seed: such a thought seemed extravagant only to those who measured God's powers by their own imagination.

The eighteenth-century preformationists had an answer to all criticisms. When the French preformationist Charles Bonnet was confronted with the argument against preformation that, if the first rabbit had enclosed within it all future rabbits, it would have had to contain $10^{10,000}$ preformed embryos, he merely answered that it was always possible, by adding zeros, to crush the imagination under the weight of numbers. He described the preformation theory as one of the most striking victories of the understanding over the imagination: 'one of the greatest triumphs of rational over sensual conviction'.

The rise of preformationist theories may have been a response to a series of philosophical problems. If matter could form organized beings, little role was left for the Divine Creator. Moreover, physical mechanisms for development, as suggested by Descartes, seemed impossible, for the mechanisms as exemplified by the laws of motion were blind, in that they were without direction. How could such forces generate the perfection of an organism? One answer, offered by the French biologist Buffon, was an interior mould which gave the embryo its form by means of 'penetrating forces'. Another proposed that the embryo contained a 'building master'. But such concepts provided no real answer at all, and they were severely criticized by preformationists such as the Swiss scientist Albrecht von Haller. The invariable production of always similar, always divinely constructed animals appears to be too great for the simple forces that produce something like a salt crystal.

By contrast, Caspar Friedrich Wolff was an ardent epigeneticist who based his ideas on two principles: the ability of plant and animal fluids to solidify and a *vis essentialis*, or vital force, which together could explain development. Wolff was a rationalist dominated by the idea of sufficient reason. He was unconcerned about the nature of his *vis essentialis*: 'It is enough for us to know that it is there, and to recognize it from its effects . . .' He objected to the preformationists' reliance on God rather than on a *cause* of generation. Much of his famous debate with Haller related to the development of blood vessels surrounding the embryo, which is a striking early event in chick development. Haller believed that the blood vessels pre-exist but only gradually become visible, whereas Wolff claimed that the blood vessels arose during development. Haller charged that Wolff was making an unwarranted assumption: namely, that if one cannot see a structure it does not exist.

One of the most attractive features of physics is that it can provide a set of basic mechanisms which can explain an extraordinary variety of phenomena. The basic mechanisms themselves may, like Newton's laws, not be easy to understand, but they are much simpler than the varied movements of objects which they can explain. A major difficulty with the preformation/epigenesis debate was that the explanations being offered were as complex as the phenomena themselves. Preformation itself was, of course, hardly an explanation, for it simply said that everything was there from the beginning, and the concepts 'building master' and 'vital force' in relation to epigenesis were no less complex than embryonic development itself.

Was there evidence that could have settled the question in favour of either epigenesis or preformation? The answer seems to be 'no', for this was a problem that lingered on for at least another century and a half. Embryology remained essentially descriptive, and there was no causal analysis. It was only at the end of the nineteenth century, with the beginnings of experimental embryology, that the preformation/epigenesis issues began to be more clearly defined and some of the issues were settled. Understanding that all organisms are made of cells was essential and put the whole problem in quite a different light, for it then became possible to realize that new cells are generated by cell multiplication during development. Advances in cell biology, aided by the invention of better microscopes, made it clear that embryos could develop by epigenesis, and eventually, only recently, it was realized that DNA provides the programme for embryonic development. Preformation died slowly, and even at the end of the last century August Weismann's theory, in which determinants in the egg nucleus were the controlling elements, was preformationist in concept.

The preformationist/epigenesis controversy in the eighteenth and nineteenth centuries can be thought of as being premature, reflecting the premature state of embryology as a science. The problem was just too difficult for the time, and advances had to be made in other areas of biology, particularly cell biology, before progress in understanding development could be made. To use again Peter Medawar's wonderful aphorism, science is the art of the soluble.

This history of embryology can offer parallels with the current status of psychoanalytic theory. Is psychoanalysis a science? Is it helpful to consider whether or not it is common sense? Is Popper's falsification criterion helpful? The issue here concerns the explanations offered by psychoanalysis for aspects of human behaviour, not its therapeutic effectiveness.

'The intention is to furnish a psychology that shall be a natural science; that is to represent psychical processes as quantitatively determinate states of specifiable material particles.' This is Sigmund Freud's opening sentence of his manuscript *Project for a Scientific Psychology* in 1895. At the end of his life, Freud insisted that his psychoanalytic enterprise had the status of a natural science, and he claimed that the explanatory gains from proposing an unconscious mind 'enabled psychology to take its place as a natural science like any other'.

Today, however, there are those who argue that Freud was guilty of 'scientific self-misunderstanding'. It is argued that the criteria and methods of the physical sciences are inappropriate in thinking about psychoanalysis and other complex aspects of human behaviour. Instead a hermeneutic reading is proposed, by which it is meant that psychoanalysis should be viewed as an interpretive procedure, rather like interpreting a written text. Narrative explanations are always dependent on the context within which they are given and are thus complexly related to all the various cultural influences that affect the context. But, as Adolf Grünbaum has cogently argued, this approach obfuscates the whole issue. The statements of psychoanalytic theory are about tendencies or likelihoods of some behaviour occurring and are indeed causal statements: they are about cause and effect. Concepts relating to the unconscious, libido, Oedipus complex, and, above all, repression have entered our everyday thinking and are used to provide causal explanations. We need to know how reliable such explanations are, for there can be no doubt that psychoanalysis has transformed the way in which we think about human behaviour.

The concept of repression of unwanted thoughts which are too painful or disgusting is central to psychoanalysis. It derives from the original ideas of Breuer and Freud in 1893, and can be considered to be the cornerstone on which the whole of the structure of psychoanalysis rests. In the course of the treatment by hypnosis of patients with hysterical symptoms, Freud and Breuer observed that there seemed to be a release from the symptoms if the patient had a cathartic experience which revealed the underlying cause. They concluded that for each distinct symptom affecting the neurotic patient, the patient had repressed the memory of a trauma that had closely preceded the onset of the symptom, and that the trauma had some analagous features with the symptoms. The famous example was Breuer's patient Anna O., who had a phobia for drinking liquids. They claimed that she had repressed the sight of a dog drinking water from a friend's glass, which had disgusted her. By recalling the incident, the repression was lifted and there was a dramatic disappearance of the symptom. This idea was developed by Freud into a model in which not just recent traumas were important but in which symptoms would, in general, reflect childhood repression with a sexual content. (Ironically, recent evidence shows that Anna O. was far from cured and had several relapses over a period of years.)

The mental apparatus was divided by Freud into an ego, an id and a super-ego, each of which was involved in controlling the flow of psychic energy. The id is that part of the unconscious mind that is governed by irrational forces such as aggression, while the ego operates rationally and the super-ego acts as the moral conscience. Thus the ego withdraws energy from all associations which are unpleasant, and this results in repression of a memory or emotion which is still stored in the unconscious. It is, apparently, no easy matter for the ego to keep unconscious thoughts under control, and so the unconscious desire to injure someone, for example, requires defensive manoeuvres which result in the desire appearing under a different guise, and such repressed desires may also come out in a dream. While some of these ideas are novel and surprising, they do have an element of common sense about them, for they essentially locate in the mind three people with behaviours and feelings with which we are all more or less familiar. Behaviour results from the conflicting attitudes each has.

When Popper came across these ideas, in around 1920, he described the partisans of psychoanalysis as seeing confirming instances everywhere: their world was full of verifications of the theory. No matter what happened, the theory was always capable of explaining it. In this continual confirmation he saw the weakness of what he regarded as their inductivist methodology. Popper thus abandoned verification as a strong basis for a scientific approach and proposed that only falsification is an important criterion for a science. For Popper, 'these . . . clinical observations which analysts naïvely believe confirm their theory cannot do this any more than the daily confirmations which astrologers find in their practice.' But does that mean that if psychoanalysis were falsifiable, or if at least parts were, it would be a science?

Grünbaum has argued that Popper's criticisms are unfair, for did not Freud modify his theories in the light of his clinical experience? For example, Freud even considered giving up the psychoanalytic method of investigation when his theory of hysteria based on seduction collapsed after he had come to the conclusion that his patients' reported seductions were fantasies. (One of the peculiarities, and weaknesses, of the psychoanalytic method is that there is usually no way of telling whether a patient's recollection of early events is fact or fantasy. This has bedevilled the whole question of claims concerning sexual abuse.) Another case relates to Freud's theory of paranoia. His hypothesis was that repressed homosexual love is necessary for someone to be afflicted by paranoid delusions. One of his patients was a young lady who thought that the man she was involved with was arranging for their lovemaking to be photographed, in order to disgrace her. In the initial psychoanalytic session, Freud could find no sign in her of a homosexual attachment, so either his theory was wrong or the young lady's report about her lover was correct. If indeed the young woman was paranoid, Freud was prepared to abandon his theory that the delusion of persecution invariably depends on homosexuality. But rather than praise Freud for accepting falsification, we should be critical of his conclusion. For how could Freud ever be sure that there was no homosexual attachment? Even after years of analysis, how could he claim the absence of such a homosexual feeling? There is nothing that requires that it had to emerge during the analysis. Thus, far from providing a good example of the refutation of psychoanalytic theories, it does just the opposite: it shows how difficult, if not impossible, it is to falsify such a theory.

Freud regarded dreams as the royal road to a knowledge of the unconscious activities of the mind and claimed that repressed infantile wishes are the causes of all dreams. In brief, when asleep, the ego's vigil on the id is relaxed and unwelcome thoughts try to enter consciousness and might disrupt sleep if they were allowed to enter. The unwelcome thoughts are paired with material retrieved from recent experience, and the true 'meaning' or 'latent content' of the dream is thus disguised – transformed – into bizarre forms and with symbolic representation. (What, one wonders, does the pairing, the transformation?) Freud himself revised his basic theory that all dreams were basically wish-fulfilment. He felt that he had satisfactorily disposed of objections to this idea based on the occurrence of anxiety dreams and punishment dreams – these, he explained, were the fulfilment not of instinctive impulses but of the censoring, critical super-ego – but, even so, there was one set of dreams he regarded as inexplicable: the dreams of victims of traumas such as war. Such victims regularly relieve their traumatic experiences in their dreams, and Freud could not see what wishful impulse could be satisfied by harking back to such exceedingly distressing experiences. But since there is no clear indication as to how a trauma is defined or on whether a wish has been fulfilled, the reliving of traumas is really no less, or more, explicable than any other aspect of Freud's theory of dreams.

The cases just quoted mimic both scientific method and science, but they barely qualify as science at all, because both the phenomena and the theory are so ill-defined. The problem with psychoanalysis is not philosophical but lies in the nature of the theory and the state of the subject: many (probably all) of the concepts in the theory are so loosely defined that the phenomena cannot be defined unequivocally and independently. Take trauma, for example. How does one decide what constitutes trauma? If seeing a dog drinking from a glass is traumatic, then there must be thousands of events in our lives which are traumatic. Could there be a trauma-free person? A further weakness is the role of repression in causing neuroses: it is apparently a necessary but not sufficient condition. What, then, provides the sufficient conditions under which repression of traumas causes clinical symptoms? Unless these are clearly specified, one is left with no theory at all: at best there is a weak correlation of poorly defined events. Even worse, the raw data of psychoanalysis are not verifiable by multiple independent observers.

Then take the concepts of id, ego and super-ego, each of which has a character almost as complex as the phenomena they are trying to describe. Add to these a concept of psychic energy, different development stages such as oral and anal – each replete with a gamut of human emotions such as disgust, anger, desire and jealousy – and then allow both positive and negative interactions. It becomes, in principle, impossible to make any meaningful predictions or explanations: the concepts are so vague that it is almost impossible to imagine any behaviour that would either verify or falsify the theory. But these ideas do in a sense enlarge our common-sense understanding of how people behave and that is why the ideas are so seductive. And one should not deny the usefulness of collecting data which might, for example, link early experiences with later behaviour patterns. It could be very important to show, for example, that the effects of early separation of a child from its parents has a serious implication for later development.

Another aspect of the unscientific nature of psychoanalysis is the presence of different 'schools' – the Freudians, the Kleinians, the Jungians and so on. One can ask how their differences could, in principle, be resolved. What evidence, what experiment, what new data would persuade one group to change their ideas? There does not seem any way of resolving the differences. In part, the problem may be that each group forms a closed system.

It is also not possible, at the present time, to do any experiment at a lower level of organization – that is, at the level of brain function or neurophysiology – which would contradict psychoanalytic theory. Current explanations of dreaming couched in neurophysiological terms or computer analogies provide no explanation of the content of specific dreams. How, then, could one show that there is or is not an id or an Oedipus complex? At present it is not possible to relate the ideas of psychoanalysis to any other body of knowledge: they are entirely self-contained.

The current situation in psychoanalysis is in some way similar to the study of embryonic development in the eighteenth century. The claims of the rival theories of preformation and epigenesis could not have been resolved at that time because the state of biological knowledge and of technology were both inadequate. It would be hard to deny that the eighteenth-century embryologists were scientists: they designed experiments and made observations to the best of their ability – their science was simply premature and primitive. Both of the rival groups had enormous difficulty in accounting for the emergence in the embryo of highly organized patterns and forms, and invoked the idea of a 'building master' or 'vital force' or just assumed that everything was preformed. These were essentially ad hoc inventions, effectively having the same complexity as the phenomena they were meant to explain. In this sense they resemble the ego, id and super-ego and the emotions assigned to the unconscious.

Those engaged in psychoanalysis are dealing with a much more difficult problem than embryonic development. Not only is the subject-matter so much more complex, it is not easily accessible to experimental investigation in the way that embryos are. It is not known what equivalent to the 'cell' is required for understanding human behaviour, or even whether such an equivalent exists. Psychoanalysis is much worse off than eighteenth-century embryology.

One should be suspicious of ideas, like those of psychoanalysis, which have been so easily incorporated into our everyday thinking. If the rest of the physical world follows laws quite different from common sense, it would be surprising if the workings of that most complex of organs, the brain, could be so readily understood. For example, recent studies show just how unnatural the workings of the brain are with respect to language. Vowels are handled in a different way to consonants, and verbs and nouns are stored in different regions, as is shown by brain damage in specific regions. Even inanimate and animate nouns are categorized.

It can be argued that human behaviour and thought will never yield to the sort of explanations that are so successful in the physical and biological sciences. To try to reduce consciousness to physics or molecular biology, for example, is, it is claimed, simply impossible. This claim is without foundation, for we just do not know what we do not know and hence what the future will bring. No matter whether analogies between computers and the brain are correct, ideas about the problems of thinking and brain function have been greatly influenced by them. A characteristic feature of science is that one often cannot make progress in one field until there has been sufficient progress in a related area. The recent advances in understanding cancer were absolutely dependent on progress in molecular biology.

Ageing is a current area in which it might be premature to try to do research. Although there has been extensive discussion about the nature of ageing – such as whether it is genetically determined and whether it reflects the accumulation of random errors in the genes and proteins of the cell – there has been remarkably little progress. Until there was progress in understanding how genes work and control the behaviour of cells, it was not even possible to begin to think about the problem in concrete terms, and even now it is very difficult to know what research programme would be appropriate. For this reason, relatively few scientists work on the problem of ageing.

Should claims for paranormal phenomena be treated in the same way? That is, should the paranormal be regarded as a premature field that will eventually prove fertile? Those who believe in paranormal phenomena like telepathy, levitation, psychokinesis (the ability to move objects by the action of the mind) and astrology claim that these involve phenomena for which contemporary science has no explanation. Moreover, they complain that conventional science totally ignores the field even though it seems to contradict mainstream ideas and could reveal quite new dimensions about human potential and behaviour.

While there are numerous reports of paranormal phenomena, they are almost without exception anecdotal. Good evidence in the presence of independent observers, preferably including a professional conjuror who could detect fraud, is simply not available for any of the phenomena. As an explanation for this, it is suggested that somehow doing a proper experiment on paranormal phenomena makes the subjects self-aware and so eliminates the phenomena. It is thus very difficult to assess the reality of such phenomena. Nevertheless there are many reports, and the question arises as to how, for example, science can deal with levitation – that is, the claim that people can rise spontaneously off the ground in the absence of any known force – or with communication between minds in the absence of any known means of transmission of information. At present such phenomena are inexplicable by science. Moreover, they seem to be so at variance with everything we know about physics that, in the absence of very persuasive evidence, it is very difficult to take them seriously. Levitation and telepathy *may* exist; the Queen *may* be a Russian spy; but both would require remarkable evidence to persuade us to give up our beliefs to the contrary. One cannot but have sympathy with Michael Faraday, who, when asked once too often to witness some new paranormal phenomenon, replied, 'I will leave the spirits to find out for themselves how they can move my attention. I am tired of them.'

Many so-called paranormal phenomena fit nicely with

Langmuir's concept of pathological science. Irving Langmuir was a distinguished chemist who, about forty years ago, coined the term 'pathological science' in an informal but now famous lecture. He focused on a number of phenomena which had startled the world of science during his career but which had subsequently faded from view. The Langmuir criteria for pathological phenomena are that the maximum effect observed is very small, near the limit of detectability; the magnitude of the effect seems independent of the cause; claims of great accuracy; usually a fantastic theory; and criticisms are met by ad hoc excuses. The mind-reading experiments associated with extra-sensory perception were popular some time ago and fit these criteria quite well. In these experiments, subjects were asked to guess the character on a card which was held by an experimenter behind a screen. The results claimed to show a result that was statistically better than chance and so implied telepathic communication. Again and again the results were subjected to statistical analysis. Some subjects were better than others. There were accusations of fraud, and in some cases (but by no means all) fraud was established. After a flurry of enthusiasm, the entire series of experiments has virtually disappeared without trace, but will no doubt surface again in another form.

Much of the evidence for the paranormal deals with apparently trivial phenomena, such as guessing the nature of hidden cards and coincidences. In some ways the presenting of such evidence implies that anyone can do science and no special training is involved. Whereas conventional scientific knowledge is obtained in a painstaking way, with breakthroughs and flashes of insight being rare events, it is characteristic of the paranormal that major 'discoveries' are easily obtained without any special knowledge. It offers a way of getting knowledge 'on the cheap'. I have only to compare how hard it is to establish in my field, embryology, even a very simple piece of knowledge – such as when the cells in the developing arm make the decision to become a humerus – and the ease with which evidence for near-miraculous events like levitation and psychokinesis appears to be established. Whereas my tiny bit of information takes many work-years, experience of levitation – even though it invokes unmeasured forces and challenges the basis of physics – can be established, and apparently accepted, in seconds.

It is this lack of requirement for scientific knowledge, together with the concept of vitalism, that links some of the paranormal and some aspects of fringe medicine. Vitalism is an idea which assigns to human life, particularly conscious-ness, a special quality which must forever remain outside conventional science. Vitalism is usually associated with an anti-reductionist stance, the view being that life cannot be reduced to mere physics and chemistry and that a more holistic approach is required. While there is a genuine prob-lem about how to relate different levels of organization – such as the atomic, chemical, cellular and organismic – to each other, and about which level is the most appropriate on which to tackle a particular set of problems, that is not what the anti-reductionists and vitalists have in mind. Any philosophy that is at its core holistic must tend to be anti-science, because it precludes studying parts of a system separ-ately – of isolating some parts and examining their behaviour without reference to everything else. If every process were dependent on its part in the whole then science could not have succeeded. We can study cells outside of the body and particular chemical reactions outside of cells – the success of biochemistry is due to just such isolation of parts – but that is not to deny the importance of also studying the systems as a whole. The holists' unwillingness to consider explaining life in terms of, for example, molecular biology and their desire to invoke some special life-force effectively restore the soul and make an afterlife possible. The paranormal also provides humans with magical forces to control their environ-ment and their health directly. Such beliefs are very seductive and need not necessarily be connected with vitalism, though they often are.

It is remarkable that so many people have been taken in by Uri Geller's claim to have special powers as shown by his ability to bend spoons. (No one seems concerned that they are always bent at the weakest point – to bend the bowl of the spoon would be a more impressive trick.) Even some sociologists of science thought this might represent revol-utionary science in the Kuhnian sense – as revolutionary as

Einstein's or Darwin's contributions, say. Those who investigated it did not know whether paranormal metal bending was 'real' or not – nor, as sociologists, did they care: they made it clear that it would make not one jot of difference to their analysis. Unfortunately, they missed the really interesting aspect, which is why so many people are taken in by these absurd claims. It really does matter from any point of view whether the bending is real or fraudulent. While it is understandable that sociologists wish to take a neutral stance in a scientific controversy, it is necessary for them to recognize that Geller's claims are quite outside science. If not, one might just as well investigate, as science, the production of rabbits out of hats and the sawing in half of ladies.

By taking a conventional scientific viewpoint, however, is there not perhaps the danger of missing out on important discoveries made by 'amateurs'? The scientific equivalent of the great artist starving, neglected, in his garret would be that of the brilliant untrained scientist working outside conventional science, perhaps in a basement, and being scorned by the Establishment. Yet the history of science provides no good example of this, in this century at least. Einstein comes close to this image, as initially he worked unknown in a Swiss patents office. But when he submitted his papers to a physics journal the editor was so impressed that he dispatched a colleague to Zurich to find out more about their remarkable author. It is also true that the geologist Wegener was treated very badly with respect to his ideas about continental drift. Even so, I am sympathetic to vigorous rejection of the absurd: completely open minds may turn out to be completely empty. That Mars is made of red cheese is a testable falsifiable hypothesis in Popper's terms. Should claims such as this be taken seriously? No, they should be rejected as absurd and are nothing to do with science.

The physicist Richard Feynman, when told a story about flying saucers, told a believer that the existence of flying saucers was not impossible, just very unlikely. His questioner claimed that Feynman was being unscientific; if he could not prove flying saucers to be impossible, how could he say they were unlikely? Feynman's reply was that it is scientific only to say what is more likely and what is less likely, and that his guess was the more reliable. Feynman's view of science was that it proceeded by informed guesses whose implications were compared with experiment.

B,

Astrology is another case where the scientist's guess that it is absurd is almost certain to be right. In astrology, the moment of birth is taken as the decisive time in the subject's life: calculations are made to find out how the planets appeared in the heavens at that moment. The subject's birth chart then reveals a pattern of 'cosmic actions', interpreting which involves assessing the various combinatorial inter-actions between the sun, the moon and nine planets, which offers an enormous number of possibilities from which to choose.

Astrology had for many centuries almost the status of a universal law. It was widely held that the heavens influenced earthly, inferior events, and such a view was subscribed to from Aristotle right through to Bacon and Kepler. By con-trast, St Augustine thought that astrology enslaved human free will and he vigorously condemned it. He used the argu-ment that twins, with virtually identical birth times, could have quite different characters. To this the astrologers replied that the twins' instants of birth were in fact different. Augus-tine's rejoinder was that if they wished to take into account such small time-intervals then the accuracy of their predic-tions was, to put it mildly, highly suspect, for he could not believe in such a high degree of accuracy. Debates like this, as so often with pseudo-science, failed to make much progress.

For example, the sixteenth-century astrologer Hieronymus Wolf predicted the date of his own death and gave away all his worldly goods when this approached. When the predicted date of his death passed without incident, he was too ashamed to reclaim his possessions and excused his error on the grounds that he had not given the position of the planet Mars sufficient consideration – showing how astrological predic-tions can always be saved from falsification. And, even if certain claims of astrology were falsifiable, that would not make it a science.

Claims have been made that distinction in certain areas is linked to star signs. Scientists, for example, have a high fre-quency of births when Saturn has risen and a low frequency when Jupiter is superior in the skies. However, correlation alone is a long way from cause and effect, and it is not at all clear what precise claims astrology actually makes. The very implausibility of a cause whereby the planets could influence our lives has been recognized since Newton and has led to a decline in astrology among serious thinkers. Unless a link can be made with the rest of science, astrology remains a pseudo-science, linked to the paranormal. And, because it is believed in by so many people worldwide, it provides another nice example of the attractiveness of magical thinking.

Scientific modes of thought are psychologically uncomfortable, whereas magic may be seen as a means of defending the self against the hostile world which is not easily given up. I cannot help but be struck by the similarity between certain paranormal beliefs and the beliefs of children that give them a magical view of the world, as described by Piaget (Chapter 1). In many cases, it will be recalled, the child is under the impression that reality can be modified by a thought and has the animistic belief that the will of one object can act on that of others. There is, in children, a magical causality according to which all things revolve about the self.

The capacity for self-delusion, even among scientists, should never be underestimated: conviction can have profound effects on observation. Marcello Malpighi, in the seventeenth century, believed the chick embryo was preformed in the egg, even though his own beautiful observations provided the best available evidence that he was wrong. Another example is the N-ray affair, which has never been satisfactorily explained.

X-rays were discovered in 1895, and other radiations from radioactive materials were identified shortly afterwards. René Blondlot, a distinguished French physicist, announced in 1903 that he had discovered yet another form of radiation, which he called the N-ray in honour of his university at Nancy. Others, too, began investigating this new phenomenon, which had remarkable properties; for example, almost all the materials that the rays passed through were opaque to ordinary light. Blondlot was even challenged by some physicists who claimed that they had been the first to discover N-rays. The number of papers published in the leading

French journal on N-rays reached a peak in 1904, but then R. W. Wood, an American physicist, visited Blondlot's laboratory and published a report in *Nature*. Wood was persuaded that N-rays did not exist. Not only did all his tests with Blondlot's apparatus fail, but he reported that the results were unaffected when he somewhat mischievously removed a key element of the apparatus. Blondlot initially went to great lengths to respond to Wood's criticism, but even his French supporters became increasingly sceptical and he refused to get involved in a definitive test proposed by a French journal.

The N-ray affair had a parallel in 1989. Jacques Benveniste, a senior immunologist working in a French government-sponsored research laboratory, published a paper in *Nature* claiming that, in effect, water had a memory. His experiment was to dilute a particular chemical so much that, according to classical chemistry, the solution would have no trace of the original substance left; and yet this hyper-dilute solution was found to have a similar biological effect as the original undiluted solution. Benveniste's claim was that the chemical had imprinted a memory on the water and so its biological activity persisted in its absence. This was regarded as a great boost to homeopathy, which is based on the idea that certain medicines become more potent the more they are diluted. Surprise, and outrage by some, was the response of the scientific community to seeing such a paper published in a most prestigious scientific journal. But part of the deal in accepting the paper had been that *Nature* itself would visit Benveniste's laboratory. The editor, together with a magician (a well-known exposer of hoaxes) and an 'expert' in scientific fraud duly arrived. They found no evidence of fraud, but Benveniste's claims could not be substantiated. Benveniste, for his part, claims that the trial was unsatisfactory and stands by his original claim. And there the matter rests. Like N-rays, it may quietly disappear, but more likely it will burst forth again with equally unpersuasive evidence as another example of Langmuir's pathological science.

Yet there may be something of a puzzle remaining. The idea of all bodies attracting each other was a remarkable imaginative leap by Newton, particularly since the attraction was not, in his theory, mediated by any other medium. So when Newton produced his theory of gravity – that bodies attracted each other at a distance with a force proportional to their masses – it was greeted with comments that you might think would be used by scientists like myself commenting on paranormal phenomena. 'It pleases some to return to occult qualities . . . but because these have become unrespectable they call them forces, changing the name . . .' wrote the great Leibniz. To Leibniz, gravitational attraction was 'a senseless occult quality . . . that it can never be cleared up, even though a spirit, not to say God himself, were endeavouring to explain it.' The Dutch scientist Christiaan Huygens was also very critical: 'That is something I would not be able to admit because I believe that I see clearly that the cause of such an attraction is not explainable by any of the principles of mechanics or of the rules of motion.'

Newton responded with typical vigour and said that Leibniz denied conclusions without taking fault with the premises; that Leibniz's arguments were founded upon metaphysical hypotheses, whereas he, Newton, was interested only in experiments. Moreover, he charged that Leibniz took refuge in emotive expressions such as 'miracles' and 'occult qualities' so that he might denigrate universal gravity. Yet, as Newton wrote,

It is inconceivable that inanimate brute matter, would without the modification of something else which is not material, operate on, and affect other matter without material contact . . . That gravity should be innate, inherent and essential to matter, so that one body may act upon another at a distance through a vacuum without the mediation of something else . . . is to me so great an absurdity that I believe no man who has in philosophical matter a competent faculty of thinking can ever fall into it.

But, he goes on: 'Gravity must be caused by an agent acting constantly according to certain law but whether this agent be material or immaterial is a question I have left to the consideration of my readers.' In his *Philosophiae Naturalis Principia Mathematica* he made his position absolutely clear: 'I have not been able to deduce from phenomena the reason of these properties and I do not feign hypotheses.'

Newton's position is fundamental to understanding the nature of science. Gravity was not an occult quality but a postulate from which testable observations could be made and which provided an economical way of explaining in a consistent and logical manner a very large number of phenomena. What was at dispute was the postulate to explain the observations, not the observations themselves, and he had, he admitted, no satisfactory explanation for gravity (Chapter 4). Darwin, too, had to make an assumption, about biological variation, for which he had no explanation but excellent evidence. In a way their situations were not that different from a perception of modern quantum mechanics: as Murray Gell-Mann, one of the founders of modern physics, says:

All of modern physics is governed by that magnificent and thoroughly confusing discipline called quantum mechanics invented more than fifty years ago. It has survived all tests. We suppose that it is exactly correct. Nobody understands it but we all know to use it and how to apply it to all problems: and so we have learned to live with the fact nobody can understand it.

It is just one of the painful aspects of science that scientists learn to live with: they recognize when understanding is absent, but at least they have the phenomena.

Unlike science, religion is based on unquestioning certainties. It is neither easy nor natural for most people to live with uncertainty, and religion can provide a solution to many problems, particularly moral ones. Thus all religious belief can be regarded as natural – in the sense that most societies, both present and past, have had religious beliefs that can provide an explanation of its members' origins and make some meaning of their lives. This presents a problem for scientists who have to reconcile their views with religion or reject it, since, as I will try to show, religion and science are incompatible. There is also another basic problem, for, as Tolstoy pointed out, 'Science is meaningless because it gives no answer to our question, the only question important for us: "What shall we do and how shall we be?" ' Tolstoy was right in that science cannot provide moral guidance.

These are problems with a long history. As we have seen

(Chapter 3), Averroës argued that the cultivation of science should be totally independent of any tenet of Muslim creed. He avoided a scientific discussion of miracles reported in the Koran: 'Of religious principles it must be said that they are divine things which surpass human understanding, but must be acknowledged although their causes are unknown.'

The tradition of Averroës was taken to its logical conclusion by David Hume, who put it succinctly: 'Our most holy religion is founded on Faith, not on reason.' Opposing the traditional doctrine that science and religion were complementary, Hume maintained that they were mutually exclusive. Religion, he argued, is not even a form of knowing: it is rather a complex kind of feeling. Believers could not legitimately employ material events or rational arguments to support their religious belief. The religious man can only be a fideist: one who believes without recourse to science or reason. For Hume, religion simply postulated unknown causes. He was opposed to natural theology – the idea that based the existence of God on the majestic and wondrous design of nature. Reason, he considered, is limited to the realm of human experience and therefore it cannot decide ultimate questions such as the origin of the cosmos: 'we have no experience of divine attributes or operations.'

Scientists have to face at least two problems that tend to drive them in opposite directions. On the one hand, however successful their theories may be, there will always be an irreducible set of laws or fundamental particles which must be taken as given, without any cause. There must come a point at which there is no cause, no explanation: the origin of the universe must ultimately be inexplicable and something must be taken as an unquestioned starting-point. Science can never provide the answers to everything: even when there is a unified theory that might explain everything, there must always be something – the justification for the theory, the basic postulates – that remains unexplained, unaccounted for, and scientists must accept this. This might drive some scientists to arguing that God provides the starting-point, and that God wound up the universe and set it going. But now the scientist is driven in the opposite direction, for postulating a God is to postulate a causal mechanism for which there is neither evidence nor any foundation – a postulate that cannot be falsified. A scientist may perhaps believe in a God, but he or she cannot use God as an explanation for natural phenomena. He escapes embodied presence and perception since he is not in space and his existence cannot be demonstrated. Thus his existence has to be of a radically different character from the reality of the world. God is in this sense a non-existent entity. How can a scientist deal with a non-existent entity?

It could be gratifying, even comforting, for the scientist to find support in religion, even though it is not compatible with scientific belief. But if there were an intellectually legitimate path from the scientific world to a religious belief in something more cosmic, God-like, there is no reason to believe that the path would lead to a benevolent Christian God, or the God of any other faith.

Yet many of the greatest scientists, from Galileo to Einstein, have had no difficulty in being deeply religious. Newton even saw himself as God's prophet and spent innumerable hours showing how the secrets of nature were hidden within the Bible. Michael Faraday's scientific creativity was intimately linked to his Christian belief. He was a member of the small Sandemanian sect, who believed in the literal interpretation of the Bible, and Faraday thought that, similarly, scientists should read the book of nature as directly as possible, through experiment, and should avoid abstract mathematical theories. For Einstein, 'A religious person is devout in the sense that he has no doubt about the significance of those suprapersonal objects and goals which neither require nor are capable of rational foundation . . . A legitimate conflict between science and religion cannot exist. Science without religion is lame, religion without science is blind.' What he is saying is perhaps similar to Tolstoy's statement quoted earlier.

This paradox may be understood in terms of the natural nature of religion compared to science. To follow up Tolstoy's point, the scientist, or anyone else, without religion has to face an indifferent chaos and has to accept that all human hopes and fears, all ecstatic joys and dreadful pains, all the creative torments of scholars, artists and saints and technicians are going to vanish forever, without trace. If, as Halévy puts it, 'Reason is insignificant as compared to the instinct by which we live,' then some scientists are able to set aside the conflict between science and religion. For being religious need not interfere with one's scientific activity and can even have a positive effect, so different are the two modes of thinking.

One approach, by a religious scientist, John Polkinghorne, is to view the theological enterprise as summed up in a phrase from St Anselm: faith seeking understanding. Theology is a reflection upon religious experience, following Whitehead's definition: 'The dogmas of religion are attempts to formulate in precise terms the truths described in the religious experience of mankind. In exactly the same way the dogmas of physical science are attempts to formulate in precise terms the truths discovered by the sense perceptions of mankind.' However, this approach begs the key question: namely, whether religious experience is of a different kind from all other experiences. Why should religious experience be treated as different from any other experience and not be subject to scientific inquiry in the normal way? However intense and remarkable religious experience may be, that in itself cannot justify it being granted a privileged autonomy. There is nothing in religious experience that is incompatible with science; the incompatibility only arises when it is claimed that religious experiences are quite different from any other and involve, for example, supernatural phenomena such as a deity or miracles. One way out of this dilemma is thought to be that religion, like the strange behaviour of subatomic particles, may call for its own 'special kind of rules for discourse'. But there is no justifiable connection.

A somewhat different perspective is to base religion, such as the Christian tradition, on historical evidence such as the Scriptures. This at once raises the problem of miracles: 'Admit the existence of God, of a personal God, and the possibility of the miraculous follows at once.' By invoking miracles it is thought possible to form a coherent picture of God's activity in the world that embraces both the fact that in our experience dead men stay dead and also that God raised Jesus on Easter Day. Science is not in a position to contradict these special cases on the basis of its generalized investigations. On this view, miracles are seen not as celestial conjuring tricks but as signs – 'insights into a deeper rationality than that normally perceived by us'. Similarly, it is argued that, just as in science the interesting anomaly can be an important lead in pointing the way forward, so the events associated with the life of Jesus, which have an anomalous character and are apparently inconsistent with science, must take into account the spiritual dimension and should be used as an example of this dimension. But Hume has already countered such arguments: 'no testimony is sufficient to establish a miracle, unless the testimony be of such a kind that its falsehood would be more miraculous than the fact which it endeavours to establish.'

Some religious scientists have argued that it is not logically valid to use science as an argument against miracles, since to believe that miracles cannot happen is as much an act of faith as to believe that they can happen. In a way they are right, since consistency and universality in the laws governing nature are basic, and usually unstated, assumptions that scientists make. But such assumptions are testable and so are quite different ones from those required by religious beliefs. Like the paranormal, the evidence for miracles or the existence of heaven, hell or an after life is not sustainable within the context of science, so scientists ought to continue to deny the possibility of miracles until presented with evidence to the contrary. Those who believe that religion and science are compatible have to believe in things demonstrably unscientific and to assert the existence of entities or processes for which no shred of evidence exists.

This analysis has thus far concentrated on the compatibility between science and religion and has avoided the direct conflict that was recognized by Averroës. Scientific evidence is in direct conflict with the Scriptures. Humans, so science claims, are closely related to the apes, and women do not come from Adam's rib. The response of a group of Christian fundamentalists has been to devote a great deal of effort to arguing that evolutionary theory suffers from serious deficiencies and that creation science, a doctrine compatible with the book of Genesis, provides a far better explanation.

The creationist campaign is an attack not merely on evolutionary theory but on the whole of science, for, if its supporters' claims about evolutionary science were to lead to it being dismissed, other major fields of science would also have to be dismissed. For example, the creationists maintain that the earth is only a few thousands of years old. If this is true, then all the estimates based on radioactivity (and hence physics itself) are false and therefore most of astronomy and geology have to be rejected. One must understand that the creationist science is Bible-based and hence linked to a set of presuppositions that cannot be altered, or proved false. Creationist science is not science if only because it precludes change in ideas; such changes are fundamental to science.

The creationists, like some of those who support the paranormal, mimic science in order to bolster their arguments. Thus the creationists lay down criteria for science and then argue that evolutionary theory does not fit these criteria. Their argument is that science requires proof and that the evidence provided by evolutionary biology does not constitute the required proof. They charge evolutionary scientists with basing their beliefs on faith, not evidence.

But, as we have seen, science is concerned not with absolute truth but providing a usable and reliable body of knowledge about the nature of the world. Change is crucial to science, but not change without good evidence. The great physicist Lord Kelvin, it will be recalled, was wrong about the age of the earth, which he too thought was not very old, because he based his calculations on the cooling of the earth; but radioactivity, an important source of heat in the earth's core, had not yet been discovered in his day.

The attack of the creationists is based in part on the claim that evolutionary theory cannot be falsified à la Popper. But, as we have seen, falsification is just one aspect of science – and, in any case, current evolutionary theory could easily be proved false. For example, if it were shown that many acquired characters were inherited, or that mammalian fossils were found in rocks whose age antedated the vertebrates, or that the DNA of birds was more similar to that of worms than to that of cats, or that animals changed rapidly without selection, the impact on current ideas in evolutionary mechanisms could be fatal.

Even though science and religion are in basic conflict, one should be cautious in assuming either a radical decline of religious belief in recent years or that any decline is due to science. Many scientists, around 50 per cent, are religious, and in the United States more than 90 per cent of the population admit to religious beliefs. Moreover, the social historian David Martin has pointed out that it is necessary to examine more than just the figures for church attendance: superstition is still strongly with us. To quote Martin on secularization in the so-called age of science:

Far from being secular our culture wobbles from being a partially absorbed Christianity, biased towards comfort and the need for confidence, to belief in fate, luck and moral governance incongruously joined together. If we add to these layers of folk religiosity the attraction of Freudianism and Marxist mechanics for segments of the intelligentsia, it is clear that whatever the difficulties of institutional religion they have little connection with any atrophy of the capacity for belief.

In his view, vast numbers of people work on two basic principles: one is the rule of chance – fate – the other is a moral balance in which wicked deeds are punished. I believe that many of us continue to subscribe to this magical, more natural, image of the world.

8

Moral and Immoral Science

Many people perceive the ethical and social implications of science as a major issue. This underpins much of the hostility that is felt towards science in some quarters. For example, nuclear weapons and genetic engineering arouse considerable anxiety and lead to questions about the wisdom of encouraging scientific investigation in all fields and about whether scientists take sufficient responsibility for their work. Scientists are seen as meddling with nature, and there is a widespread feeling that scientists are so blinkered by their research and so motivated to make new discoveries that any experiment that can be done will be done, no matter what its implications. The image of scientists as so many Dr Frankensteins looms large. Newspapers repeatedly print stories with headlines warning against the dangers of genetic engineering and the human genome project, coupled with the cliché of 'scientists playing at God'. Of course these anxieties coexist with the hope that science will provide the solution to major illnesses such as cancer and heart disease and to genetic disabilities like cystic fibrosis.

Some of these anxieties have an ancient history and are linked to the idea that knowledge is dangerous. Prometheus was punished for bringing knowledge to the world, and Faust for wanting it too much. Lest one thinks that the biblical tree of knowledge, for the tasting of the fruit of which man was expelled from Eden, was only about the knowledge of good and evil, Milton's version in *Paradise Lost* makes the issue clear. The serpent addresses the tree as 'Mother of Science' and Adam tells the Archangel Raphael that, while his thirst for knowledge has been largely satisfied by what Raphael has told him about the Creation, some doubts do remain. Raphael's response is rather patronizing: he doesn't blame Adam for asking, but

> . . . the Great Architect
> Did wisely to conceal and not divulge
> His secrets, to be scanned by them who ought
> Rather admire.

God is, he says, rather amused by their 'quaint opinions'. What, Raphael asks, does it matter if the sun be the centre of the world? His advice is: 'Solicit not thy thoughts with matters and . . . Be lowly wise. Think only what concerns thee and thy being.' In Francis Bacon's time it was perceived that 'knowledge puffeth up', and it has even been suggested that Francis Bacon's task, and main achievement, was to show that science was not after all Mephistophelean.

The issues can best be analysed along two main lines. First I will attempt to determine just what responsibilities scientists have: what obligations they have as scientists as distinct from their responsibilities as citizens. My suggested obligations are only that they must inform the public about the possible implications of their work and, particularly where sensitive social issues arise, they must be clear about the reliability of their studies. My other line of analysis is related to the first and necessitates examining to what extent ignorance about the nature of science and its conflation with technology have contributed towards presenting a misleading representation of the role of science. For the applications of science are not necessarily the responsibility of scientists. Moreover, many apparently new ethical issues are in fact old ones that have become confused because they are linked with a science that is strange and new, such as genetic engineering. In order to justify these statements I will first consider some aspects of the development of the atomic bomb, particularly from the viewpoint of one scientist, since it illuminates some of the moral issues involved. Moreover it is a moral tale. Then the history of eugenics will be used as an immoral tale.

In 1933 *The Times* quoted the physicist Lord Rutherford, who had just split the atom, as saying that 'anyone who looked for a source of power in the transformation of atoms was talking "moonshine".' A Hungarian physicist, Leo

Szilard, was staying at the Imperial Hotel, Bloomsbury, and read the article. He was reminded of H. G. Wells's *The World Set Free*, published as long ago as 1914, in which both the development of atomic energy and an atomic bomb are described. To Szilard, pronouncements of experts to the effect that something cannot be done were always irritating. As he later wrote, 'This sort of set me pondering as I was walking the streets of London, and I remembered I stopped for a red light at the intersection of Southampton Row . . . I was pondering whether Lord Rutherford might not be proved wrong.' It was at that instant that the idea of a neutron chain reaction came to him. This was a crucial moment in the history of the atomic bomb. While he could not see at that moment just how one would go about finding an element that would give a chain reaction, or what experiments would be needed, the idea never left him. He was convinced that in certain circumstances it might be possible to set up a nuclear chain reaction and so liberate energy on an industrial scale, and construct atomic bombs.

When Szilard took his ideas to British physicists, he found no support. Rutherford virtually threw him out of the office, and another physicist told him he would have no luck with such fantastic ideas in England: perhaps, it was suggested, he should try Russia.

Nevertheless Szilard stuck to his idea, and in 1934 he applied for a patent which described the laws of a chain reaction. Because of his reading of H. G. Wells, he did not want the patent to become public and possibly used by the Germans, and so he assigned it to what, I would guess, was a rather puzzled British Admiralty. By 1936 his own and others' experiments had extinguished his faith in the possibility of a chain reaction, and so he wrote to the Admiralty waiving the secrecy on the patent and suggested that it be withdrawn altogether. But in 1938, now living in the USA, he learned that uranium had just the properties that might sustain a chain reaction. He now tried to persuade his physicist colleagues not to speak publicly about the possibility of a chain reaction, as this might give invaluable information to the Germans, who could use it to build an atomic bomb. But the Italian physicist Enrico Fermi would not take him seriously, because he thought the possibility of a chain reaction was still unlikely. Other physicists, like Nils Bohr, could not accept secrecy in physics, as it was completely against the openness of science. Bohr was also not convinced of the likelihood of producing a nuclear explosion. Fermi and Szilard nevertheless hesitated as to whether to publish their own results, which made a chain reaction seem very likely, but they were pre-empted by a publication in *Nature* which effectively made public the same conclusion.

Szilard now contacted Einstein and persuaded him to write the famous letter to President Roosevelt which was sent on 2 August 1939: 'Sir, some recent work by E. Fermi and L. Szilard, which has been communicated to me in manuscript, leads me to expect that the element uranium may be turned into a new and important source of energy in the immediate future . . . This new phenomenon would also lead to the construction of bombs . . .' Einstein asked the President that some permanent contact be maintained between the administration and the group of physicists working on chain reactions in America and that funds be provided to speed up experimental work. In May 1940, President Roosevelt spoke to the Pan American Scientific Congress in Washington. Germany had just invaded Belgium and Holland. He told them that if the scientists in the free countries would not make weapons to defend their freedom, then freedom would be lost. He assured them that it was not the scientists of the world who would be responsible. In effect he gave the scientists a presidential exoneration for the consequences of any weapons that they would help construct.

Meanwhile a committee had been set up in Britain to look at the possibility of a chain reaction bomb, and in 1941 it reached the conclusion that it would be possible to make an effective uranium bomb. On 9 October 1941 the British report was taken to Roosevelt and influenced the decision to proceed. It was at this meeting that the policy on the future of the bomb and any future decisions were moved firmly under the President's control. As Richard Rhodes, from whose book *The Making of the Atom Bomb* much of this story has been taken, writes:

From this time on, a scientist could choose to help or not to help build nuclear weapons. That was his only choice. The surrender of any further authority in the matter was the price of admission to what would grow to be a separate, secret state with separate sovereignty linked to the public state through the person and by the sole authority of the President.

The commitment to build an atomic bomb was made by Roosevelt alone.

Szilard remained in Chicago while the bomb was being developed at Los Alamos, New Mexico. In March 1945 he began to examine the wisdom of testing bombs and using bombs. It became clear to him that the war against Germany would soon end, and so he began to question himself about the purpose of continuing the development of the bomb, and about how the bomb would be used if the war with Japan had not ended by the time the USA had the first bombs. The initial motivation of getting ahead of the Germans was no longer there. He drafted a memorandum for Roosevelt: he saw little point in approaching anyone else. He again persuaded Einstein to write to the President. Einstein did so, pointing out that the terms of secrecy did not permit Szilard to give him information about the events in question, but he did emphasize Szilard's concern about the lack of adequate contact between the scientists who were doing the work and the members of the administration who formulated policy.

Szilard argued that, by preparing to test and use atomic bombs, the United States was moving along a road leading to the destruction of the strong moral position it had hitherto occupied in the world. When, Szilard argued, other countries acquired nuclear weapons, the US military supremacy would be lost and an arms race would begin. He went on to consider the possibility of international control rather than an American monopoly of the atomic bomb.

Roosevelt died in May 1945, and it was Truman's Secretary of State, James Byrnes, who met with Szilard. He argued that Congress would want results for its $2 billion investment, and that not to test was not an option. Also, the USA's having a bomb might make the Russians 'more manageable'. The bomb was inevitably and successfully tested on 15 July.

It could be regarded as a triumph of engineering. Many, many scientists and engineers were involved. The technology was amazing, but at heart it was merely a gigantic superstructure that made Szilard's original idea work.

Even before the bomb was tested, Szilard was circulating among the scientists working on the bomb a petition to present to Roosevelt's successor, President Truman. It started, 'Discoveries of which the people of the United States are not aware may affect the welfare of this nation in the near future.' It continued by arguing against the use of the bomb now that there was no danger of the enemy using it against the United States:

... a nation which set the precedent of using these newly liberated forces of nature for purposes of destruction may have to bear the responsibility of opening the door to an era of devastation on an unimaginable scale ... We, the undersigned, respectfully petition first, that you exercise your power as Commander-in-Chief to rule that the United States shall not resort to the use of atomic bombs in this war unless the terms which will be imposed upon Japan have been made public in detail and Japan knowing these terms has refused to surrender ...

Sixty-seven scientists signed the petition, but it never reached the President.

One of those who refused to sign the petition was Edward Teller, who wrote to Szilard: 'First of all let me say that I have no hope of clearing my conscience. The things we are working on are so terrible that no amount of protesting or fiddling with politics will save our souls ... Our only hope is in getting the facts of our results before the people.'

The bomb was dropped on Hiroshima on 6 August 1945.

There are several lessons to be learned from this tale. First there is no clear relation between ideas and implementation, between science and technology. Building of the bomb was a technological commitment, and its achievement was based on scientific knowledge. To the very end there was no certainty that it would work as planned. The gap between basic scientific knowledge and application was in this case enormous. The principles were well founded, but their application was a gigantic engineering feat which had little to do with science in the sense that it provided no new understanding of the way in which the natural world works.

In emphasizing the technology, I do not mean to underplay the science involved. This can be illustrated in relation to the problem as to why the Germans didn't build the bomb. Werner Heisenberg and other German physicists may have missed some crucial scientific point, which may account for their failure to build an atomic bomb. Thus Heisenberg's statement, made after the war, that German physicists were spared the decision as to whether or not they should aim at producing atom bombs is as likely to mean that they didn't know how to as to mean that Hitler showed no interest. After the defeat at Stalingrad a decision was taken not to invest heavily in nuclear weapons but to concentrate on rockets.

Secondly, the decision to build the bomb was a political and not a scientific decision. It is not uninteresting to speculate what might have been the course of history if Szilard had not persuaded Einstein to write his first letter to Roosevelt. The bomb would probably not have been built during the war – it would have come too late, and building it in peacetime might not have been politically or economically possible. The scientists involved saw a clear demarcation between their responsibility and that of government, which Robert Oppenheimer, who was in charge of building the bomb, made explicit:

The scientist is not responsible for the laws of nature, but it is the scientist's job to find out how these laws operate. It is the scientist's job to find the ways in which these laws can serve the human will. However, it is not the scientist's responsibility to determine whether a hydrogen bomb should be used. That responsibility rests with the American people and their chosen representatives.

Szilard's behaviour illustrates a third lesson: one of the most important obligations to emerge from this tale is that of openness, exemplified by his emphasis after the war on telling the public about the implications of scientific knowledge. It is true that Szilard argued for secrecy before the war, but it is also clear that it is not really possible to block the advance of knowledge. In general, all great discoveries made by particular scientists would sooner or later have been made by someone else. It is Szilard's later emphasis on public involvement, unless national security is threatened, that we should focus on. The necessity for the public to be informed about science and its implications is a major obligation for scientists.

There is another aspect to the bomb that needs to be put in perspective which shows how the alienation and misunderstanding of science makes it seem more culpable. The number of deaths at Hiroshima from the effects of the bomb was about 200,000 compared with the 100,000 deaths in Tokyo due to firebombing earlier in 1945 and a similar figure for Dresden. These figures must be seen in the light of the 100 million people killed by man-made means this century. About half of these, that is 50 million, were killed by guns or conventional bombs, and the rest by privation such as labour camps, displacements and man-made famine. No one associates these deaths with science, because the technology involved was simple and understandable.

Pulling a trigger is easy. There was no temptation to blame science for the 50 million deaths from guns or conventional bombs. Certainly these were part of the war machine, but the technology was based on gunpowder, which is more familiar than nuclear weapons. Nuclear weapons, however, are alienating, because most of us do not understand nuclear physics. While not denying the disastrous potential of nuclear weapons, we should not underestimate this alienation, for I believe it lies at the core of many of the so-called problems about the social responsibility or irresponsibility of science. It leads to confusion between the weapon and crime, particularly since the weapon itself is so mysterious. It thus becomes the duty of all scientists to minimize this alienation whenever possible. Only in this way may it be possible to dissuade people from seeing the creation and use of death machines as the responsibility of scientists. They have in this respect no more responsibility than other citizens. Those who regard the scientists in America who helped develop the atomic bomb as immoral should consider a scenario in which the scientists had decided not to do so and Germany had in fact been successful. Would they have been satisfied that such a decision should have been taken for them by the scientists?

Since scientists are providers of knowledge, they have an obligation to report the implications of that knowledge; but the implementation, the application, of that knowledge is a social and political decision which it is not for them to take. In these terms, science is not responsible for misapplication of knowledge. But how, then, can we give credit to science for its positive applications? The answer, I think, is that knowledge, in the scientific sense, is intrinsically good. All understanding of our world is a positive achievement, and science can be applauded for this – even more so when it leads to positive applications ranging from penicillin to power generation. But is all knowledge beautiful and neutral in the sense I have suggested? The story of scientists and eugenics raises some difficult questions.

In 1883, Darwin's cousin Francis Galton coined the word 'eugenics'. It came from the Greek 'good in birth' or 'noble in heredity'. Eugenics was defined as the science of improving the human stock by giving 'the more suitable races or strains of blood a better chance of prevailing speedily over the less suitable'. For Galton, science and progress were almost inseparable. Men could be improved by scientific methods, in the way that plant breeders improve their stock. Would it not, he wondered, be 'quite practicable to produce a highly gifted race of men by judicious marriages during several consecutive generations'? The scientific assumptions behind this are explicit: most human attributes are inherited.

Galton's views were derived from ideas about natural selection and evolution: 'The processes of evolution are in constant and ponderous activity, some towards the bad and some towards the good. Our part is to watch for opportunities to intervene by checking the former and giving free play to the latter.' Not only was talent perceived of as being inherited, so too were pauperism, insanity and any kind of perceived feeble-mindedness. Darwin himself was reported by Wallace to be gloomy about the future of humanity, for he thought that those 'who succeed in the race for wealth are by no means the best or the most intelligent, and it is notorious that our population is more largely renewed in each generation from the lower than from the middle and upper classes.'

These ideas, amplified by Karl Pearson from University

College London, received support from a variety of sources, which included Fabians such as Bernard Shaw and psychologists like Havelock Ellis.

An American, Charles Davenport, was particularly influenced by the idea of eugenics, and in 1904 he persuaded the Carnegie Foundation to set up the Cold Spring Harbor Laboratories for the study of human evolution. From his studies on human pedigrees, Davenport came to believe that certain races were feeble-minded. Negroes were biologically inferior; Poles were perceived of as independent and self-reliant, though clannish; Italians tended to crimes of personal violence. He expected the American population to become, through immigration, 'darker in pigmentation, smaller in stature, more mercurial . . . more given to crimes of larceny, kidnapping, assault, murder, rape and sex-immorality'. His aim was to promote negative eugenics – preventing proliferation of the bad. To this end he favoured a selective immigration policy to prevent the contamination of the 'germ-plasm' (the genetic information transmitted from parents to offspring) from without, and to deal with 'badness' within the present population he tried to prevent reproduction of those whom he considered genetically defective. He was in fact pursuing a policy put forward by a Timothy Nurse in England over 200 years earlier: a gentleman 'ought be as careful of his race as he is of his horses, where the fairest and most beautiful are made choice of for breed'. Davenport even remarked that it would be a progressive revolution if 'human matings could be placed upon the same high plane as that of horse breeding'.

Davenport's approach to human genetics was in terms of the action of single genes, though he knew of polygenic inheritance – that is, a character being determined by several or many genes – like, for example, skin colour. Thus he suggested that prostitution was due to an 'innate eroticism'. Yet Davenport saw himself as a scientist who wished to base his ideas on sound investigations. Thousands of Americans were persuaded to fill out their 'Record of Family Traits'. In 1907 a national Eugenics Education Society was set up in England, and others were formed in America. Though membership was small, the influence of these societies was large, and in the 1920s Fitter Families contests took place at the Kansas Free Fair. The American Society even had a Eugenics Catechism:

QUESTION: What is the most precious thing in the world?
ANSWER: The human germ-plasm.

A further sense of the feeling of the time is given by the contest sponsored by the American Eugenics Society for essays on the decline of 'Nordic fertility'.

Much of this information comes from Daniel Kevles's book *In the Name of Eugenics*. As Kevles points out, the geneticists warmed to their priestly role, and the list of distinguished scientists that initially gave eugenics positive support is depressingly impressive: Fisher, Haldane, Huxley, Castle, Morgan. According to the American geneticist Herbert Jennings in his 1930 book *The Biological Basis of Human Nature*, the world is to be operated on scientific principles. The conduct of life and society are to be based, as they should be, on sound biological maxims.

One approach to negative eugenics was sterilization to prevent contamination of the germ-plasm. It is estimated that between 1907 and 1928 about 9,000 people were so treated in the United States, all of them classified under the general rubric of 'feeble-minded'. In a famous court case in 1927, Justice Oliver Wendel Holmes gave a judgement in favour of sterilization which included the statement that 'the principle that sustains compulsory vaccination is broad enough to cover cutting the Fallopian tubes . . . Three generations of imbeciles is enough.'

In the 1930s, Huxley, Haldane, Hogben, Jennings and other biologists at last began to react against many of the wilder claims for eugenics. But it was too late, for the ideas had permeated into mainland Europe. As the geneticist Benno Müller-Hill wrote in his book *Murderous Science*, 'The ideology of the National Socialists can be put very simply. They claimed that there is a biological basis for the diversity of mankind. What makes a Jew a Jew, a Gypsy a Gypsy, an asocial individual asocial, and the mentally abnormal abnormal, is in their blood, that is to say in their genes.' It is hard not to believe that this was based on the genetic ideas of the eugenic movement. For example, Professor Fischer, a Professor of Anthropology and Rector of the University of Berlin, a promulgator of such views, was asked by Davenport in 1929 to become Chairman of the Committee on Racial Crosses of the International Federation of Eugenics Organizations. Thus it is quite easy to see a direct line from the eugenics movement to the statement by the most famous animal behaviourist Konrad Lorenz:

It must be the duty of racial hygiene to be attentive to a more severe elimination of morally inferior human beings than is the case today . . . We should literally replace all factors responsible for selection in a natural free life . . . In prehistoric times of humanity, selection for endurance, heroism, social usefulness etc. was made solely by hostile outside factors. This role must be assumed by a human organization; otherwise humanity will, for lack of selective forces, be annihilated by the degenerative phenomena that accompany domestication.

Another metaphor from Lorenz is the 'analogy between bodies and malignant tumours on the one hand, and a nation and individuals within it who have become asocial because of their defective constitution'.

In 1933, Hitler's cabinet promulgated a Eugenic Sterilization Law which can be considered as leading directly to the atrocities by doctors and others in the concentration camps. This law made sterilization compulsory for anyone who suffered from a perceived heredity weakness, including conditions from schizophrenia to blindness.

Doctors, in general, are not scientists: they are technologists, more like engineers, applying knowledge of human biology. What we have to consider is the responsibility of the scientists who effectively laid the foundation for the genetic theories that underlay the cruder versions of biological determinism – namely, that genetic factors determine complex human behaviour patterns. One cannot dismiss them solely by saying they were bad scientists, for the question of biological determinism – that many human characters like aggression and altruism are genetically programmed – is very much still with us, in the form of sociobiology.

With the wisdom of hindsight, we may feel smug about

how misguided many of the ideas fuelling the eugenics movement were. But, for all we know, many of its supporters were in other respects good and honourable scientists: they were just wrong. They were bad scientists in terms of both the science they did and their obligations.

The scientists who promoted the views of the eugenics movements may have been honourable with respect to their science. They could, perhaps, plead genuine ignorance or fault in dealing with the data, but they completely failed to examine critically the social implications of their conclusions and to make them public. In fact, much more culpably, their conclusions seem to have been driven by what they saw as the desirable social implications, and they totally failed to inform the public about the likely reliability of their conclusions.

Scientists have the obligation to examine the social implications of their work, not in order to decide how or if it should be used – that, as in the case of the bomb, is a political decision – but in order to make clear the reliability of the interpretations of the observations. In some areas of science it matters little to the public whether a particular theory is flawed, or even wrong. It matters mainly to the scientists if, say, some of the current ideas about how embryos develop turn out to be wrong. By contrast, it matters a great deal if, as in the case of human genetics, complex behavioural characteristics are treated as being controlled by genes and behavioural influences are ignored. Scientists have an obligation to make the reliability of their views in these sensitive social areas clear to the point of overcautiousness. And the public should, wherever possible, demand the evidence and critically evaluate it.

What is at issue is how scientists should handle delicate issues like the relationship between race and intelligence. Is research into such areas legitimate? Or are there, as the literary critic George Steiner has argued that there are, 'certain orders of truth which would infect the marrow of politics and would poison beyond all cure the already tense relations between social classes and ethnic communities'? In short, he asks whether there are doors immediately in front of current research which are marked 'too dangerous to open'. Provided the obligations to examine and explain the social implications and to make clear the reliability are fulfilled, my answer to Steiner is a cautious 'no'. The main reason is that the better the understanding we have of the world, the better the chance we have to make a just society. It is to the credit of the biological community that the debate on the validity of applying to human behaviour ideas of sociobiology derived from studies of animals has been carried on with great vigour.

When we think about the social responsibility of scientists, we are not primarily concerned with the natural duties of all citizens in our society, such as to help one another, not to inflict unnecessary suffering and so on. According to a contemporary moral philosopher, John Rawls, these duties apply to us all, without regard to certain voluntary choices, such as choice of career, that we have made. By contrast, specific obligations result from our having made a particular choice, such as marrying or standing for public office. So the question is, what obligations above natural duty do scientists, as distinct from other citizens, have to society? To what extent does the privileged knowledge that scientists have entail additional obligations? The issues are not essentially ethical ones, for temptation to immoral acts related to science does not seem to present a special problem, though scientists must not, of course, steal ideas, be fraudulent or fail to take due care with experiments on animals, for example.

It may be useful, rather, for scientists to make use of Rawls's first principle, outlined below, when they put forward ideas that have social implications for others – ideas like those suggested by some sociobiologists which encourage ideas about social determinism: that, for example, class structures are socially inevitable, that aggressive impulses towards strangers are part of our evolutionary heritage and that there are basic and ineradicable differences between the sexes that doom women's hope for genuine equality. They should imagine themselves, following Rawls, in the original position in which the rules for society are being set up, but with none of the parties knowing his or her place in society – there is a veil of ignorance. The parties do not know their sex, their natural abilities or even the generation to which they belong. They are brought together to agree detailed rules based on a general principle of justice. All social values – liberty and opportunity, income and wealth, and the bases of self-respect – are to be distributed equally, unless an unequal distribution of any or all of these values is to everyone's advantage. The scientists should then consider the reliability they would give to their conclusions in such a situation. Would someone who believes, with deep conviction, that his or her research demonstrates the natural differences of women, or of certain races, still maintain that conviction? For it is quite likely that in this imagined situation he or she might belong to one of those groups. The response might, surprisingly, be 'yes', for this could well lead to the preferential distribution of, say, jobs or money to disadvantaged groups, to compensate them at the expense of the advantaged groups if it were established that they were naturally disadvantaged. In this situation, knowledge about presumed inherited differences could be used to design a more just society. A less sensitive example than race is how to teach children of mixed abilities, some of which are directly attributable to genetic differences.

But we do not live in such an idealized situation. In our real world, conclusions about the biological bases of human attributes and racial differences can feed current prejudices and have a severe negative influence. If any social measures are to be based on scientific knowledge, then the reliability of that knowledge must be made very clear. This is no small problem, but that does not remove the obligation.

Many people may still not be persuaded by the sharp distinction I draw between scientific knowledge and its application, between science and technology: that doctors and engineers are not necessarily scientists. This distinction is based not on purity or snobbery, but on implementation of knowledge which may be based on science. Nevertheless, there are areas where the boundaries may not at first appear to be very sharp, as in the case of the application of genetic engineering and gene therapy.

Genetic engineering provides the means for altering the genetic constitution of animals and plants. It offers great hope for solving problems relating to pest control, the excessive use of fertilizers, energy use and a host of other areas. Yet the very term 'genetic engineering' conjures up fears about tampering with nature. These fears have a long tradition. According to Greek myth, Poseidon made King Minos' wife fall in love with a white bull and the result of their union was a monster – the minotaur. In more recent times we have had Mary Shelley's Dr Frankenstein and H. G. Wells's Dr Moreau: both created monsters and fed deep-rooted fears about chimeras. This tradition certainly has given biologists involved in any sort of genetic engineering a bad image.

To focus on a specific issue, it is now possible to insert genes into human cells to correct genetic defects. Should this be done? A distinction must be drawn between insertion of genes into somatic or body cells, which will not be passed on to future generations, and the introduction of genes into germline cells, eggs or sperm, which will result in the genes being passed on to all subsequent generations. The insertion of genes into body cells could provide a powerful technique for correction of major genetic diseases such as sickle-cell anaemia, cystic fibrosis, muscular dystrophy, thalassemia and many others. Four thousand disabilities due to single-gene defects are known.

Why, then, does the prospect of gene therapy of this sort seem to pose major ethical issues? The transferring of genes into body cells does not raise any new ethical issue, since the introduction of new genetic material is routinely done with organ and bone-marrow transplants. Moreover, treatments with radiation and certain drugs themselves alter the genetic material. The safety problems, even if modified viruses are used as carriers to insert the germs, seem to be no different to those involved in any new medical treatment. All new medical interventions carry risks, and there are well-established procedures for introducing new drugs, for example. It is hard to see how anyone could object to curing disabling genetic diseases. The anxieties must rest on the dangers of stepping on to the 'slippery slope'.

The slippery slope is where one joins Drs Frankenstein and Moreau. Once there is the possibility of introducing genes into body cells to treat disease and disability, then it is inevitable, so the argument goes, that doctors and scientists will insert genes to change the character of people. Genes could be inserted that would make people taller, handsomer, cleverer and perhaps even happier. And, the argument continues, even if the technique were limited initially to body cells, sooner or later genes would be introduced into the germline.

But what objection is there to manipulating the genes of somatic cells so as to provide the recipient with what is perceived as some highly desirable quality? Is it any different from cosmetic surgery? Imagine that it were possible by genetic engineering to improve, say, memory or some particular mental skill during an individual's lifetime. It might be thought that this would be abused, since only those with sufficient money might have access to the treatment. Such a procedure might give some an unfair advantage. There would also be problems of safety, of ensuring that there were no undesirable side-effects. While adults could have a choice, treating children does present an ethical issue. But are these issues really any different to what happens now? Parents give an advantage to their children by giving them private lessons, by sending them to the very best schools. At present, self-improvement is a highly valued social attribute. Jogging may improve the cardiovascular system, bran the bowels, and meditation the whole body. What, then, is wrong with the 'supermarket' approach in which genes would be available at a price and with suitable warnings about possible side-effects? Why should adults be denied chemicals that may improve our well-being? Individual responsibility and choice, provided the results harm no one else, is fundamental to a democratic society. We might find it distasteful that someone might use gene therapy to change themselves in a particular way, but perhaps it is something we have to live with.

The possibility in the future of introducing genes into germline cells does raise special problems, because the effects will be passed on. But how different is the choice of introducing genes into the germline from parents choosing to have children knowing that they have a high chance of their child having a genetic defect? If women have the right to bear children infected with AIDS, why cannot mothers choose to have a genetic defect in their eggs corrected? It seems almost immoral not to do so. For to correct the genetic abnormality in this way is to correct it for the descendants forever, and this would lead both to fewer children inheriting the abnormality and to fewer abortions. But the case for introducing genes into the human germline is fraught with dangers. To do so is to make a change for future generations, and the possible chance of things going wrong makes the procedure so risky as to make it, at present, unacceptable. It is banned by law in many countries, including most of Europe and the US. Moreover, prenatal diagnosis and abortion is a preferable approach for genetic diseases and, in the long run, makes germline therapy unnecessary.

It is at first sight curious how concerned people are about genetic engineering, which has so far damaged no one. By contrast, smoking, AIDS, drugs and alcohol have caused massive damage to children *in utero*. Perhaps a clue to this attitude lies, again, in the unnatural nature of science. It stems from the fear of the unknown – of processes, words, techniques that people do not understand. Compare attitudes to genetic engineering with those to euthanasia. Both have ethical aspects and are of public concern, but no one links the euthanasia debate with science, and the reason is that it involves no complex science or technology which they do not understand. It is worth remembering the tremendous hostility there was to vaccination in the last century; it was only when the public had sufficient understanding of it that it became accepted and became part of 'common sense'. A strong case be made for trying to make people 'DNA-literate' so that they can appreciate the issues associated with genetic engineering; only then may many of the misplaced fears disappear.

There is, for example, considerable anxiety about the human genome project – a project which aims to map all human genes on the chromosome and even to determine the nature, and ultimately the function, of every gene. Many are frightened by the detailed information about human make-up that this project will provide. It will certainly provide improved means for the early detection of genetic 'abnormalities' such as a predisposition to cancer or heart disease or mental ill-health, and this could be used by employers or insurance companies in antisocial ways. But, against this, it must be understood that the genetics of such disabilities is already being worked out by classical techniques, and surely it is better to have the possibility of the individual knowing about any problems early on, rather than waiting for the disease or abnormality to become evident. The human genome project would provide an early indication of genetic characters by enabling an early examination of the DNA, rather than waiting until the effect is expressed. For example, Huntington's Chorea is a tragic and incurable neurological disturbance which affects men in their fifties. Its genetics is well understood, and an examination of the DNA of an individual who is at risk can now show whether he will be affected or not. This might seem to pose a major ethical issue: should people be tested and be told the result, particularly if it is positive, knowing that nothing can be done. But that is the wrong way to look at the problem: rather, one should ask what people who might develop the disease actually want. The results are unequivocal: they prefer to have the test available.

I am not trying to suggest that genetic engineering does not raise any difficult issues, but I am suggesting that most of the problems are ones that have been met before. For example, the problem of knowledge of potential disposition to an illness is important in relation to insurance companies. But insurance companies already have to face this problem in relation to other illnesses, such as AIDS and smoking. There is a quite different set of problems with respect to safety when genetically engineered organisms are released into the environment, but safety issues with respect to the environment are hardly new. One should not muddy one's appreciation of these problems because of ignorance about genetics or a primitive fear of mythical chimeric animals. What is essential is openness and public debate.

These are issues not for the scientist but for the public at large. For the scientist who has special access to knowledge about genetics, for example, the issue is whether the genes will bring about the hoped-for changes and what dangers there might be. Even for the introduction of genes into human cells, it is not for the scientists or the doctors to decide on the wisdom or otherwise of such procedures: the obligation of the scientist and the doctor is simply to spell out the procedures' implications.

It is not for scientists to take moral or ethical decisions on their own: they have neither the right nor any special skills in this area. There is, in fact, a grave danger in asking scientists to be more socially responsible – the history of eugenics alone illustrates at least some of the dangers. Asking scientists to be socially responsible, other than by being cautious in areas where there are social implications, would implicitly be to give power to a group who are neither trained nor competent to exert it.

Scientists will undoubtedly be faced with difficult social and ethical problems in areas as diverse as nuclear power, ecology, clinical trials and research on human embryos. In each case their obligation, in addition to those responsibilities of every citizen, is to inform the public and to be open. For example, the issues relating to research on human embryos are complex, and biologists have much to contribute to issues such as at what stage the developing embryo can be regarded as an individual. But there are many other issues outside their specific area of competence, including consideration of the rights of the foetus and whether it is ethically permissible to use a patient to do research without that person's permission.

To those who doubt whether the public or the politicians are capable of taking the correct decisions, I would commend the words of Thomas Jefferson: 'I know no safe depository of the ultimate powers of the society but the people themselves, and if we think them not enlightened enough to exercise that control with a wholesome discretion, the remedy is not to take it from them, but to inform their discretion.' When the public are gene-literate, the problems of genetic engineering will seem no different in principle from those like euthanasia and abortion which are not obfuscated by an alienating scientific ignorance.

With significant exceptions, I believe that the scientific community has, on the whole, behaved responsibly with respect to the public. It would be a great error if sole responsibility for ethical decisions were given to scientists or if they were to assume it, for these are decisions that belong to the public as a whole – decisions that are essentially social and political. No one would expect scientists to be responsible for deciding whether abortion should be legal or not, though scientific information would be vital. The decisions ultimately have to be made by our elected representatives, informed by the best available scientific knowledge.

It is important to remember that, as the French poet Paul Valéry said, 'We enter the future backwards.' Scientists cannot know all the technical or social implications of their work. There was no way that those who were investigating the peculiar behaviour of certain bacteria with respect to the bacterial virus that infected them could know that they would discover restriction enzymes which cut DNA at specific sites and which are now fundamental to genetic engineering. Today's moonshine is tomorrow's technology, and it is with technology and politics that the real responsibility lies. Even so, one must guard against taking scientific ideas as dogma and treating science as infallible.

9
Science and the Public

Pp. 172-178

Ⅰ

.A,

If science is so unnatural and leads to misunderstandings about science and even hostility towards it by some of the public, what can be done? The question is important, because science provides the best way of understanding the world. The achievements of unifying the laws of physics and of synthesizing new chemicals are breathtaking, and there is every reason to believe that the future achievements in biology will be equally impressive. Yet the misunderstandings remain, even though making one's work accessible to the public is at last becoming acceptable within the scientific community. No longer is such 'popularization' treated by scientists with contempt and suspicion, as if it were a vulgar activity. The hope is, of course, that if the public understand science better they will be in a better position to understand its role in current life and will be better able to make informed decisions on issues relating to the environment, genetic engineering, nuclear power and many other concerns. Also, it is felt that if the public have an improved understanding of science they will have a more sympathetic attitude towards it. However, attempts at 'popularization' perhaps failed to emphasize two important features of science: what science cannot do, the problems that cannot be solved by science, and, of course, its unnatural nature.

When Václav Havel, quoted in the Introduction, talks about science being the sole legitimate arbiter of all relevant truth, he does both science and truth a disservice. He has also forgotten Tolstoy's correct claim that science does not tell us how to live: that it has nothing to contribute on moral issues. It is the politicians, lawyers, philosophers and finally all citizens who have to decide what sort of society we will live in. It is necessary constantly to remind Havel, and the many like him, that knowledge is not the same as its applications. To blame science for the bomb or for industrial pollution is to fail to realize that the decisions involved are political and economic, and not just about scientific understanding, and so is unfair. To blame science may be convenient, but it is wasted effort.

It is true that science might have killed God for some people,˙ but many scientists are filled with religion, and the capacity for mystical belief still seems very large for many people. One need only look at the extraordinary popularity of astrology for evidence of this. Scientific knowledge and method may be uncomfortable, but such discomfort is surely better than ignorance. And, while it can in no way tell us how to live, science may help us achieve specific aims once these have been chosen. Science could be used to alleviate genetic diseases by genetic engineering if society as a whole finds it acceptable. If not, then, like euthanasia, it can be banned. These are decisions which everyone must help take: certainly it would be a great folly to entrust decisions about how to use science to scientists or any other group of experts.

It is not just moral and political issues for which science may be unable to provide solutions: science may not be able to provide solutions to all technical problems. Indeed, it is not really possible to predict what science will produce in the future – it is not possible to predict radical innovations. While one may be able to predict inventions based on current knowledge, such as a cure for cancer based on current technologies, we have no idea what future scientific advances will bring; that is of the very nature of scientific advance.

Dostoevsky feared that science could predict the future:

Therefore all there is left to do is to discover these laws and man will no longer be responsible for his acts. Life will be really easy for him then. All human acts will be listed on something like logarithm tables, say up to 108,000 and transferred to a timetable . . . They will carry detailed calculations and exact forecasts of everything to come, and so no adventure and no action will remain.

This fear is quite without foundation. Science tries to

understand how the diversity of the physical and natural world can be explained by a limited number of laws. The phenomena to be explained are much more complex than the laws themselves. Newton's laws of motion, for example, are quite simple compared to the enormous variety of motions they can account for. It is important to realize that knowing the laws does not mean that the behaviour of the system can be described. For example, one can write down the equations of motion of three bodies which attract each other with a gravitational force, but to solve these equations so that one can actually have a description of their motion is enormously difficult, and has yet to be achieved.

To take another example, predicting global climate change is very difficult and at present far from reliable. The system is enormously complex and the models give no more than quite crude approximations. Much more basic research is required. Confident detailed predictions should be treated with caution.

Just how difficult it can be to predict the future behaviour of some systems, such as the weather, has become clear from recent studies on chaos. The basic idea is that some systems are so sensitive to the smallest of perturbations that the beat of a butterfly's wings in an English garden could lead to a hurricane in some distant part of the world. Another example which illustrates some of the problems of predictability is provided by the effect of a single electron at the end of the universe.

Consider a box containing molecules of a simple gas. Say I were to tell you the position and velocity of every one of these molecules, exactly. Using simple mechanics, you could then work out the future behaviour of the system, again exactly, as the molecules bump into one another and the walls of the box. Now introduce a small degree of uncertainty, such as a force originating outside the box and acting on the molecules. The force can be very small, say one single electron. To minimize the effect, we can put this electron a very long way away – at the end of the universe. The only uncertainty is where it is placed. The effect of this uncertain force on the behaviour of our molecules is that in less than fifty collisions for each molecule our predictions become totally wrong. Predictability lasts for less than one-millionth of a second. This clearly shows that causality is present in such systems but that detailed predictability is impossible.

Even in cases where all the facts are known, it still may not be possible to make a logical scientific choice between various possibilities. It is quite easy for a group of people to agree to choose between alternatives: a simple majority vote is the obvious way. But what if there are three or more possibilities? One might think that if everyone ranks their choice then it would easily be possible to select the most favoured choice. Consider eleven people with choices A, B, C, who order their preferences. Suppose four prefer ABC, five prefer BCA, two prefer CAB. As can be seen, A beats B (six to five) B beats C (nine to two) and C beats A (seven to four). Thus A beats B beats C beats A. And so this method can lead to a contradiction and cannot be used. In more general terms this is known as Arrow's Impossibility Theorem in economics, which says that there may be no rational solution for distributing resources among people or groups with conflicting demands. This has important implications, for it means that even when we have total information we cannot solve an important problem. So, while science can help define the problems in allocating resources in the health services, say, there need be no unique solution and compromises may have to be made.

The same is true for moral and political problems: there is no way of achieving all the desirable virtues of a 'perfect' society. For example, the philosopher Isaiah Berlin makes this clear in relation to the ideal of freedom:

One freedom may abort another; one freedom may obstruct or fail to create conditions which make other freedoms, or a larger degree of freedom, or freedom for more persons, possible; . . . the freedom of the individual or the group may not be fully compatible with a full degree of participation in a common life, with its demands for cooperation, solidarity, fraternity. But beyond all these there is an acuter issue: the paramount need to satisfy the claims of other no less ultimate values: justice, happiness, love, the realization of capacities to create new things and experiences and ideas, the discovery of truth.

We must resist being seduced by science into thinking that all problems will yield to its approach. They may, in the future. But at present our understanding of such complex systems as human behaviour and society as a whole is so limited that knowledge in these fields is barely at the stage of a primitive science. Marxism should serve as a reminder as to how dangerous 'scientific' claims to understanding social processes can be. Economic predictions, too, are remarkably unreliable. And this presents a real problem, for, as the economist Robert Heilbroner has pointed out, 'The human psyche can tolerate a great deal of prospective misery, but it cannot bear the thought that the future is beyond all power of anticipation.'

Science is wrongly perceived to be homogeneous. The anonymity of scientists, as presented by the media, has helped to contribute to the idea that scientists know everything in science – that biologists, for example, will have a good understanding of physics, and vice versa. But in fact science is quite difficult even for scientists. Physicists may have little understanding of even the basic ideas of cell biology, and biologists, on the whole, are out of their depth with much of modern physics. Even mathematicians would have to work for many months in order to understand work in a different area. And that is what makes scientists different: they feel confident that, given the effort and time, they could understand most other areas of science, if not in detail then at least in general principles. Non-scientists do not in general have this confidence, nor do they have any familiarity with scientific thinking. For example, only about 5 per cent of Americans have been found to be reasonably scientifically literate, even though about half the bills before Congress involve either science or technology.

This general lack of familiarity with scientific thinking is very clear in relation to the experience of illness, where the implications are highly relevant. Patients suffer an illness whereas doctors treat a disease – the gap between these perceptual frameworks can be big. People who are ill have an overwhelming need to make sense of their personal misfortune. Cancer patients often ascribe their condition to fate, and this may be scientifically quite close to the truth, since cancer is due to the accumulation in a single cell of a number of rare events. But fate is not seen like that, but rather as some higher power controlling our destiny. Stress and diet are widely seen as sole causes of illness, since, after all, these are the variables everyone is familiar with in their daily life. However, much current thinking about disease is in cellular and molecular terms, so explanations to a patient are very difficult. For most people, even the distinction between infections due to viruses (which are not free-living but must enter cells) and bacteria (which *are* cells) is poorly grasped.

There is no easy road to understanding science, the more so because there is no formula for scientific method. The best and probably the only way to understand science is to do scientific research, but that clearly is not an option to improve public understanding. However, it may well be that science education should take into account the unnatural nature of the subject. Instead of teaching science only as a rigorous, self-contained subject, it may be beneficial to compare common-sense ideas about the world with scientific views. Studies (Chapter 1) already show that children do better at science if they acquire some understanding of independent variables, for example. But perhaps that is not enough. They need to appreciate just how different scientific thinking is and how much more natural were Aristotle's ideas as compared to those of Galileo and Newton.

Most education in science avoids personalities. It also excludes insight into the process of science. Much biology is learned at school (and even at university) in the same way as one learns irregular verbs in a foreign language – by rote. Little would be lost if less science were taught but some insight were gained into the processes of science. Learning about creativity in science, with an emphasis on psychic courage and failure, may well be very much more valuable than some of the science itself.

Whether or not non-scientists like it, science is part of our culture. Most people's views are in some way influenced by scientific ideas – that the earth goes round the sun, that genetic defects cause disease, that radioactivity can be dangerous – even if they have a very poor understanding of the validity or basis of the ideas. Understanding the processes of science and scientific ideas is hard. Ironically, some of the ideas that have been popularized, like chaos and the peculiar features of quantum mechanics, have been used in a magical way in novels such as Ian McEwan's *The Child in Time.* We need to find ways in which to make science less alien, exciting, but not mystical. Somehow we must find a way to remove the humiliating wound to their intellectual self-esteem that non-scientists like Lionel Trilling (page xi) experience by not understanding science. This is a central problem to be faced at all levels in the education system.

Science is bound to play a central role in our lives. It is to science and technology that we shall have to look for help to get us out of some of the mess in which we now all find ourselves – a mess that involves both environmental pollution and overpopulation. Of course not all the solutions will be science-based, but science can make a crucial contribution. I cannot offer specific solutions, for the nature of discovery precludes that, but knowing how the world works may be an essential requirement for helping to save it.

Finally, we should always remember the origin of science in Greece. Even though we do not understand why it should have had its origins there, the Greek commitment to free and critical discussion was essential for science to flourish. And the same is true today. While at present science flourishes, we must be aware how easily it might wither: witness the disastrous effects Lysenko's dogmas, supported by the state, had on Soviet genetics. Those who dislike the ideas of science and think they have had a malevolent effect on our spiritual life should realize that once one rejects understanding and chooses dogma and ignorance, not only science but democracy itself is threatened. Science is one of humankind's greatest and most beautiful achievements and for its continuation, free and critical discussion, with no political interference, is as essential today as it was in Ionia.

References

Key references are marked with an asterisk.

1: Unnatural Thoughts

* Holton, G., and Brush, S. G. (1985) *Introduction to concepts and theories in physical science*. Princeton University Press, Princeton, New Jersey.

Toulmin, S. (1961), *Foresight and understanding: an enquiry into the aims of science*. Indiana University Press, Bloomington

MOTION

* Gentner, D. and Stevens, A. L. (eds.) (1983), *Mental models*. L. Erlbaum, Hillsdale, NJ

Sorabji, R. (ed.), (1987) *Philoponus and the rejection of Aristotelian Science*. Duckworth, London

CHILDREN

* Adey, P. (1987), 'Science develops logical thinking – doesn't it?' Part 1. *School Science Review*, 68, 622–30

* Driver, R. (1988), 'Restructuring the science curriculum: some implications of studies on learning for curriculum development'. In *Innovation in Science and Technology*, D. Layton (ed.), pp. 59–84. UNESCO

Kuhn, D., Amsel, E. and O'Loughlin, M. (1988), *The development of scientific thinking skills*. Academic Press, San Diego

Kagan, J. (1984), *The nature of the child*. Basic Books, New York.

Leslie, A. M. and Keeble, S. (1987), 'Do six-months old infants perceive causality?' *Cognition* 25, 265–88

* Piaget, J. (1929), *The child's conception of the world*. Routledge & Kegan Paul, London

COMMON-SENSE THINKING

* Bannister, D. and Fansela, F. (1971), *Inquiring man: the theory of personal constructs*. Penguin, Middlesex

Furnham, A. F. (1988), *Lay theories*. Pergamon Press, Oxford

* Johnson-Laird, P. N. (1988), *The computer and the mind*. Fontana, London

* Kahneman, D., Slovic, P. and Tversky, A. (1982), *Judgement under uncertainty: heuristics and biases*. Cambridge University Press, New York

Kuhn, D. (1991), *The skills of argument*. Cambridge University Press, Cambridge

* Matlin, M. W. (1989), *Cognition*. Holt, Rinehart & Winston, New York

* Paulos, J. A. (1990), *Innumeracy*. Penguin, Middlesex.

di Sessa, A. (1988), 'Knowledge in pieces.' In *Constructivism in the Computer Age*, G. Forman and P. B. Pufall (eds.), pp. 49–70. Lawrence Erblaum: Hillsdale, NJ

Waldrop, M. M. (1987), 'Causality, structure and common sense.' *Science*, 237, 1297–9

Wason, P. C. and Johnson-Laird, P. N. (1972), *Psychology of reasoning: structure and content*. Harvard University Press, Cambridge, Mass.

2: Technology is not Science

* Basalla, G. (1988), *The evolution of technology*. Cambridge University Press, Cambridge

* Boorstin, D. J. (1984), *The Discoverers*. Random House, New York

* Ferguson, E. S. (1977), 'The mind's eye: nonverbal thought in technology'. *Science*, 197, 829

Heyman, J. (1987), 'The structural analysis of gothic architecture.' *Proc. Royal Institution of Great Britain*, 59, 215–26.

* Lévi-Strauss, C. (1966) *The savage mind*. Weidenfeld & Nicolson, London

Toulmin, S. and Goodfield, J. (1965), *The architecture of matter*. Penguin, Middlesex

3: Thales's Leap: West and East

* Galileo, G. (1953), *Dialogues concerning two new sciences*. S. Drake (ed.) University of California Press, Berkeley

* Kuhn, T. (1957), *The Copernican revolution*. Harvard University Press, Cambridge, Mass.

* Lloyd, G. E. R. (1987), *The revolution of wisdom: studies in the claims and practice of ancient Greek science*. University of California Press, Berkeley

Neugebauer, D. (1979), *A history of ancient mathematical astronomy*. Springer Verlag, Berlin

Sarton, G. (1959), *A History of Science.* Harvard University Press, Cambridge, Mass.
Weber, M. (1968), *Economy and society.* G. Roth and C. Wittick (eds.) Bedminster Press, New York

CHINA
Graham, A. C. (1989) *Disputers of the Tao philosophical argument in ancient China.* Open Court, La Salle, Ill.
* Needham, J. (1969), *The grand titration: science and society in east and west.* Allen & Unwin, London

RELIGION
Funkenstein, A. (1986), *Theology and the scientific imagination.* Princeton University Press, New Jersey
* Jaki, S. L. (1986), *Science and creation.* Scottish Academic Press, Edinburgh
* Thomas, K. (1971), *Religion and the decline of magic.* Weidenfeld & Nicolson, London
Willey, B. (1934), *The seventeenth-century background.* Chatto & Windus, London

PRIMITIVE SOCIETIES
* Ashley Montagu (1974), *Coming into being among the Australian aborigines.* Routledge & Kegan Paul, London
* Horton, B. (1967), 'African traditional thought and Western science.' *Africa*, 38, 50–71 and 153–87

EVOLUTIONARY CONTINGENCY
Gould, S. J. (1989), *Wonderful Life.* Norton, New York

4: Creativity

Borkowski, J. C. and Peck, V. A. (1986), 'Causes and consequences of metamemory in gifted children.' In *Conceptions of Giftedness*, R. J. Sternberg and J. Davidson (eds.), pp. 182–200. Cambridge University Press, New York
* Cohen, I. B. (1980), *The Newtonian revolution.* Cambridge University Press, Cambridge
Cole, J. R. and Cole, S. (1972), 'The Ortega hypothesis.' *Science*, 178, 368–75
Davidson, J. (1986), 'The role of insight in giftedness.' In *Conceptions of Giftedness*, R. J. Sternberg and J. Davidson (eds.), pp. 201–22. Cambridge University Press, New York.
Feynman, R. (1965), *The character of physical law.* MIT Press, Cambridge, Mass.
De Groot, A. D. (1969), *Thought and choice in chess.* Mouton, The Hague

Gruber, H. E. (1986), 'The self-construction of the extra-ordinary.' In *Conceptions of Giftedness*, R. J. Sternberg and J. Davidson (eds.), pp. 247–63. Cambridge University Press, New York

Medawar, P. (1984), *Pluto's republic*. Oxford University Press, Oxford

* Merton, R. K. (1973), 'Singletons and multiples in science.' In *The Sociology of Science*, N. W. Storer (ed.), pp. 343–70. University of Chicago Press, Chicago

* Miller, A. I. (1992), 'Scientific creativity: a comparative study of Henri Poincaré and Albert Einstein.' *Creativity Research Journal* (in press)

Sayers, D. L. (1927), *Unnatural death*. Gollancz, London

Simonton, D. K. (1988), *Scientific genius: a psychology of science*. Cambridge University Press, New York

* Weisberg, C. (1986), *Creativity: genius and other myths*. Freeman, New York

* Wolpert, L. and Richards, A. (1988), *A Passion for Science*. Oxford University Press, Oxford [interviews with Brenner, Crick and Zeeman]

COMPUTER MODELS

Csikszentmihalyi, M. (1988), 'Motivation and creativity: towards a synthesis of structural and energistic approaches to cognition.' *New Ideas in Psychol.*, 6, 159–76

* Langley, P., Simon, H. D. and Zytkow, J. M. (1987), *Scientific discovery*. MIT Press, Cambridge, Mass.

DARWIN

* Mayr, E. (1982), *The growth of biological thought*. Harvard University Press, Cambridge, Mass.

* Gruber, H. (1981), *Darwin on man: a psychological study of scientific creativity*. University of Chicago Press, Chicago.

MOLECULAR BIOLOGY

* Hilts, P. J. (1982), *Scientific temperaments*. Simon & Schuster, New York

Jacob, F. (1988), *The statue within*. Basic Books, New York

* Judson, H. F. (1979), *The eighth day of creation: the makers of the revolution in biology*. Jonathan Cape, London

SERENDIPITY

Kohn, A. (1989), *Fortune or failure*. Blackwell, Oxford

* Roberts, R. M. (1989), *Serendipity*. Wiley, New York

Maurois, A. (1959), *The life of Sir Alexander Fleming*. Jonathan Cape, London

5: Competition, Cooperation and Commitment

* Brush, S. G. (1989), 'Prediction and theory evaluation: the case of light bending.' *Science*, 246, 1124–9

Djerassi, C. (1990), *Cantor's dilemma*. Macdonald, London

Gaston, J. (1973), *Originality and competition in science*. University of Chicago Press, Chicago

* Hallam, A. (1983), *Great geological controversies*. Oxford University Press, Oxford

* Holton, G. (1978), *The scientific imagination*. Cambridge University Press, Cambridge

Hull, D. L. (1988), *Science as a process: an evolutionary account of the social and conceptual development of science*. University of Chicago Press, Chicago

Keller, E. F. (1983), *A feeling for the organism*. Freeman, New York [a biography of McClintock]

* Kuhn, T. S. (1970), *The structure of scientific revolutions*. University of Chicago Press, Chicago

* Lightman, A. and Gingerich, O. (1992), 'When do anomalies begin?' *Science*, 255, 690–95

* Merton, R. K. (1973) 'The normative structure of science.' In *The sociology of science*, N. W. Storer (ed.), pp. 267–78, University of Chicago Press, Chicago

Mitroff, I. I. (1974), *The subjective side of science: a philosophical inquiry into the psychology of the Apollo moon scientists*. Elsevier, New York

Racker, E. (1989), 'A view of misconduct in science.' *Nature*, 339, 91–3

Shapin, S. and Schaffer, S. (1985), *Leviathan and the air-pump: Hobbes, Boyle and the experimental life*. Princeton University Press, NJ

Toulmin, S. E. (1967), 'The evolutionary development of natural science.' *American Scientist*, 55, 456–71

* Wolpert, L. and Richards, A. (1988), *A passion for science*. Oxford University Press, Oxford [interviews with Epstein and Salam]

6: Philosophical Doubts or Relativism Rampant

* Barrow, J. D. (1988), *The world within the world*. Clarendon, Oxford [the nature of physics]

Gooding D. (1990), *Experiment and the making of meaning*. Kluver, Dordrecht [Faraday's method]

* Holton, G. (1978), *The scientific imagination: case studies*. Cambridge University Press, Cambridge

Merton, R. (1973), *The sociology of science: theoretical and empirical investigations*. N. W. Storer (ed.), University of Chicago Press, Chicago

Root-Bernstein, R. S. (1989), *Discovering: inventing and solving problems at the frontiers of scientific knowledge*. Harvard University Press, Cambridge, Mass.

Weber, M. (1968), *On charisma and institution building*. S. N. Eisenstadt (ed), University of Chicago Press, Chicago

Ziman, J. (1968), *Public knowledge: an essay concerning the social dimension of science*. Cambridge University Press, Cambridge

PHILOSOPHY OF SCIENCE

Bernard, C. (1927), *An introduction to experimental medicine*. Macmillan, New York

Magee, B. (1973), *Popper*. Fontana, London

* Newton-Smith, W. H. (1981), *The rationality of science*. Routledge, London [good introduction to the problems]

Popper, K. R. (1963), *Conjectures and refutations*. Routledge, London

Putnam, H. (1987), *The many faces of realism*. Open Court, La Salle, Ill.

Rorty, R. (1991), *Objectivity, relativism and truth*. Cambridge University Press, Cambridge

Ziman, J. (1984), *An introduction to science studies: the philosophies and social aspects of science and technology*. Cambridge University Press, Cambridge

RELATIVISTIC SOCIOLOGY

* Barnes, B. (1985), *About science*. Blackwell, Oxford

Barnes, B. and Edge, D. (eds.) (1982), *Science in context: readings in the sociology of science*. Open University Press, Milton Keynes

Gross, A. G. (1990) *The rhetoric of science*. Harvard University Press, Cambridge, Mass.

Hesse, M. (1980), *Revolutions and reconstructions in the philosophy of science*. Harvester Press, Brighton

* Hollis, M. and Lukes, S. (1982), *Rationality and relativism*. Blackwell, Oxford

Latour, B. (1987), *Science in action: how to follow scientists and engineers through society*. Open University Press, Milton Keynes

* Pickering, A. (1984), *Constructing quarks: a sociological history of particle physics*. Edinburgh University Press, Edinburgh

* Woolgar, S. (1988), *Science: the very idea*. Horwood, Chichester & Tavistock, London

Ziman, J. (1985), 'Deconstructing the physical world.' *Minerva*, pp. 517–22 [review of Pickering's book]

MENDEL

* Brannigan, A. (1981), *The social basis of scientific discoveries*. Cambridge University Press, Cambridge

Jacob, F. (1974), *The logic of living systems: a history of heredity*. Allen Lane, London

* Mayr, E. (1982), *The growth of biological thought*. Harvard University Press, Cambridge

PHRENOLOGY

* Cantor, G. (1975), 'The Edinburgh phrenology debate 1803–1828.' *Annals Science*, 32, 195–218

* Shapin, S. (1979), 'The politics of observation: cerebral anatomy and social interests in the Edinburgh phrenology disputes.' In *On the margins of science: the social construction of rejected knowledge*, R. Wallis (ed.), pp. 130–38. Sociological Review Monogr. 27, Keele University, Keele

SCIENTIFIC CONTROVERSIES

Close, F. (1991), *Too hot to handle: the race for cold fusion*. Princeton University Press, NJ

* Franks, F. (1982) *Polywater*. MIT Press, Cambridge, Mass.

ANTHROPOLOGY

Douglas, M. (1980), *Evans-Pritchard*. Fontana, London

Leach, E. (1982), *Social anthropology*. Fontana, London

Lévi-Strauss, C. (1968), *Structural anthropology*. Penguin, Middlesex

O'Keefe, D. (1982), *Stolen lightning: the social theory of magic*. Martin Robertson, Oxford

7: Non-Science

Gardner, H. (1987), *The mind's new science*. Basic Books, New York

Jay, P. (ed.) (1981), *The Greek anthology and other ancient epigrams*. Penguin, Middlesex

Medawar, P. (1963), 'Is the scientific paper a fraud?' *Listener*, 12 September

EMBRYOLOGY

Cole, F. J. (1930), *Early theories of sexual generation*. Clarendon Press, Oxford

Roe, S. A. (1981), *Matter, life and generation: 18th-century*

embryology and the Haller-Wolff Debate. Cambridge University Press, Cambridge
* Wolpert, L. (1991), *The triumph of the embryo*. Oxford University Press, Oxford

PSYCHOANALYSIS
Farrell, B. A. (1981), *The standing of psychoanalysis*. Oxford University Press, Oxford
* Grünbaum, A. (1984), *The foundations of psychoanalysis*. University of California Press, Berkeley
Grünbaum, A. (1986), 'Précis of the foundations of psychoanalysis'. *Behav. & Brain Science*, 9, 21 [includes commentary by others]
Hobson, A. (1988), 'Psychoanalytic dream theory: a critique based upon modern neurophysiology.' In *Mind, psychoanalysis and science*, P. Clark and C. Wright (eds.), pp. 277–308. Blackwell, Oxford
Rycroft, C. (1981), *The innocence of dreams*. Oxford University Press, Oxford
Warrington, E. K. and McCarthy, R. A. (1987), 'Categories of knowledge: further fractionations and an attempted integration.' *Brain*, 110, 1273–96

PARANORMAL
Gardner, M. (1983), *Science: good, bad and bogus*. Oxford University Press, Oxford
Gauquelin, M. (1983), *The truth about astrology*. Hutchinson, London
Inglis, B. (1985), *The paranormal: an encylopedia of psychic phenomena*. Granada, London
* Klotz, I. M. (1980), 'The N-ray affair'. *Scientific American*, May, 122
Skrabanek, P. (1986), 'Demarcation of the absurd.' *Lancet*, 1, 960–61
Stent, G. (1972), 'Prematurity and uniqueness in scientific discovery.' *Scientific American*, December, 84–93

RELIGION
Berry, R. J. (1986) 'What to believe about miracles.' *Nature*, 322, 321–2
Brooke, J. H. (1991), *Science and religion: some historical perspectives*. Cambridge University Press, Cambridge
Cantor, G. (1991), *Michael Faraday: Sandemanian and scientist*. Macmillan, London
Houghton, J. (1988), *Does God play dice?* Inter-Varsity Press

* Kitcher, P. (1982), *Abusing science*. MIT Press, Cambridge, Mass. [analysis of creationism]
* Kolakowski, L. (1982), *Religion*. Fontana, London
Martin, D. (1967), *A sociology of English religion*. SCM Press, London
Polkinghorne, J. (1986), *One world*. SPCK, London
Withnow, R. (1985), 'Science and the sacred.' In *The sacred in a secular age*, P. E. Hammond (ed.), pp. 187–203. University of California Press, Berkeley

8: Moral and Immoral Science

Elliot, G. (1912) *Twentieth-century book of the dead*. Scribner, New York
Glover, J. (1984), *What sort of people should there be?* Penguin, Middlesex
* Kevles, D. J. (1985), *In the name of eugenics*. University of California Press, Berkeley
Kitcher, P. (1985), *Vaulting ambition*. MIT Press, Cambridge, Mass. [an analysis of sociobiology]
Lancet (1989), 'Gene Therapy.' 1, 193–4
Müller-Hill, B. (1988), *Murderous science*. Oxford University Press, Oxford
Rawls, J. (1972), *A theory of justice*. Oxford University Press, Oxford
* Rhodes, R. (1986), *The making of the atomic bomb*. Simon & Schuster, New York
Weart, S. R. and Szilard, G. W. (1978), *Leo Szilard: his version of the facts*. MIT Press, Cambridge, Mass.

9: Science and the Public

Asimov, J. (1983), 'Popularizing science.' *Nature*, 306, 119
Berlin, I. (1968), *Four essays on liberty*. Oxford University Press, Oxford
CIBA Foundation (1987), *Communicating science to the public*. Wiley, London
Durant, J. (1990), 'Copernicus and Conan Doyle: or why we should care about the public understanding of science.' *Sci. Publ. Affairs*, 5, 7–22
Durant, J. R., Evans, G. A. and Thomas, G. P. (1989), 'The public understanding of science.' *Nature*, 340, 11
Fitzpatrick, R., Hinton, J., Newman, S., Scambler, G. and

Thompson, J. (eds.) (1984), *The experience of illness*. Tavistock, London

Gillick, M. R. (1985), 'Common-sense models of health and disease.' *New England J. Med.*, 313, 700–703

Havel, V. (1987), *Living in the truth*. Faber, London

Heilbroner, R. (1991), 'Reflections. Economic predictions.' *New Yorker*, 8 July, 70–7

May, R. M. (1983), 'Preference and paradox.' *Nature* 303, 16–17

Trilling, L. (1973), *Mind in the modern world*. Viking, New York

Index